Status and Understanding of Groundwater Quality in the Monterey Bay and Salinas Valley Basins, 2005: California GAMA Priority Basin Project

By Justin T. Kulongoski and Kenneth Belitz

A Product of the California Groundwater Ambient Monitoring and Assessment (GAMA) Program

Prepared in cooperation with the California State Water Resources Control Board

Scientific Investigations Report 2011–5058

U.S. Department of the Interior
U.S. Geological Survey

U.S. Department of the Interior
KEN SALAZAR, Secretary

U.S. Geological Survey
Marcia K. McNutt, Director

U.S. Geological Survey, Reston, Virginia: 2011

For more information on the USGS—the Federal source for science about the Earth, its natural and living resources, natural hazards, and the environment, visit http://www.usgs.gov or call 1–888–ASK–USGS.

For an overview of USGS information products, including maps, imagery, and publications, visit http://www.usgs.gov/pubprod

To order this and other USGS information products, visit http://store.usgs.gov

Suggested citation:
Kulongoski, J.T., and Belitz, Kenneth, 2011, Status and understanding of groundwater quality in the Monterey Bay and Salinas Valley Basins, 2005—California GAMA Priority Basin Project: U.S. Geological Survey Scientific Investigations Report 2011–5058, 84 p.

Contents

Abstract 1
Introduction 2
    Purpose and Scope 4
Description of Monterey Bay and Salinas Valley Basins Study Unit 4
    Santa Cruz Study Area 8
    Monterey Bay Study Area 8
    Salinas Valley Study Area 10
    Paso Robles Study Area 11
    Hydrogeologic Setting 11
Methods 12
    Relative-Concentrations and Water-Quality Benchmarks 12
    Datasets for Status Assessment 13
        U.S. Geological Survey Grid Wells 13
        California Department of Public Health Grid Wells 15
        Additional Data Used for Spatially Weighted Calculation 15
        Selection of Constituents for Additional Evaluation 17
    Calculation of Aquifer-Scale Proportions 21
    Understanding-Assessment Methods 22
        Statistical Analysis 22
    Potential Explanatory Factors 22
        Land Use 22
        Well Depth and Depth to Top-of-Perforation 25
        Normalized Position of Wells along Flowpath 25
        Groundwater Age 25
        Geochemical Condition 25
    Correlations Between Explanatory Factors 25
Status and Understanding of Water Quality 35
    Inorganic Constituents 35
        Trace Elements 35
            Understanding Assessment for Molybdenum 44
            Understanding Assessment for Arsenic 44
            Understanding Assessment for Boron 48
            Understanding Assessment for Manganese and Iron 48
        Radioactive Constituents 48
        Nutrients 49
            Understanding Assessment for Nitrate 49
        Major and Minor Ions 49
            Understanding Assessment for Total Dissolved Solids 49
            Understanding Assessment for Sulfate 51
    Organic Constituents 52
        Solvents 54

Contents—Continued

Status and Understanding of Water Quality—Continued

    Organic Constituents—Continued

        Gasoline Additives 54

        Trihalomethanes 54

        Other Organic Compounds 54

        Herbicides and Fumigants 54

            Understanding Assessment for Simazine 54

        Insecticides 55

    Special-Interest Constituents 55

Summary 56

Acknowledgments 56

References 56

Appendix A. Ancillary Datasets 61

Appendix B. Use of Data From the California Department of Public Health (CDPH) Database.....75

Appendix C. Calculation of Aquifer-Scale Proportions 80

Appendix D. Calculating Total Dissolved Solids 81

Appendix E. Comparison of California Department of Public Health and
U.S. Geological Survey-GAMA Data 82

Figures

Figure 1. Map of the Monterey Bay and Salinas Valley Basins study unit, California
GAMA Priority Basin Project, and the California hydrogeologic provinces 3

Figure 2. Map showing the geographic features and study areas of the Monterey Bay
and Salinas Valley Basins study unit, California GAMA Priority Basin Project 5

Figure 3. Map showing the geologic formations and areal distribution of USGS grid and
understanding wells sampled in the Monterey Bay and Salinas Valley Basins
study unit, California GAMA Priority Basin Project 6

Figure 4. Map showing the locations of study area grid cells, U.S. Geological Survey
(USGS) grid and understanding wells, and California Department of Public
Health (CDPH) wells, Monterey Bay and Salinas Valley Basins study unit,
California GAMA Priority Basin Project, July–October 2005 9

Figure 5. Ternary diagrams showing percentage of urban, agricultural, and natural
land use in the study unit and study areas, and the area surrounding each
grid and understanding well in the Monterey Bay and Salinas Valley Basins
study unit, California GAMA Priority Basin Project 23

Figure 6. Map of land use in the Monterey Bay and Salinas Valley Basins study unit,
California GAMA Priority Basin Project 24

Figure 7. Boxplots showing well depths, depths to top-of-perforation, and perforation
lengths for grid and understanding wells, Monterey Bay and Salinas Valley
Basins study unit, California GAMA Priority Basin Project 26

Figures—Continued

Figure 8. Map showing normalized position of wells along the Salinas Valley flowpath, and conceptual model of the aquifer system in the Salinas Valley for the Monterey Bay and Salinas Valley Basins study unit, California GAMA Priority Basin Project 27

Figure 9. Boxplots and bar chart showing relation of groundwater age classification to depth to top-of-perforations, well depth, and age classification, in relation to the depth of well perforations, Monterey Bay and Salinas Valley Basins study unit, California GAMA Priority Basin Project 29

Figure 10. Graph showing relation of oxidation-reduction condition to normalized position of wells along a flowpath, and depth of perforated interval of wells, Monterey Bay and Salinas Valley Basins study unit, California GAMA Priority Basin Project 31

Figure 11. Map showing pH levels in U.S. Geological Survey (USGS) wells and California Department of Public Health (CDPH) wells, and graph showing pH plotted as a function of well depth and groundwater age classification, Monterey Bay and Salinas Valley Basins study unit, California GAMA Priority Basin Project 32

Figure 12. Graph showing maximum relative-concentration of constituents detected in grid wells, by constituent class, Monterey Bay and Salinas Valley Basins study unit, California GAMA Priority Basin Project 36

Figure 13. Graph showing relative-concentrations of gross alpha radioactivity, arsenic, boron, molybdenum, nitrate, and sulfate with health-based benchmarks, and chloride, iron, manganese, and total dissolved solids with aesthetic benchmarks in USGS and CDPH grid wells, Monterey Bay and Salinas Valley Basins study unit, California GAMA Priority Basin Project 39

Figure 14. Maps showing relative-concentrations of selected inorganic constituents for U.S. Geological Survey USGS-grid and USGS-understanding wells and for the period (July 17, 2002–July 18, 2005) from the California Department of Public Health (CDPH) database), Monterey Bay and Salinas Valley Basins study unit, California GAMA Priority Basin Project 40

Figure 15. Plots showing arsenic concentration relative to classifications of groundwater age, and well depth, manganese and iron concentrations, and pH in grid and understanding wells sampled for the Monterey Bay and Salinas Valley Basins study unit, California GAMA Priority Basin Project 45

Figure 16. Plots showing nitrate, as nitrogen, concentrations relative to classifications of groundwater age, and depth to top of perforations, classification of groundwater age, and land use, in USGS-grid and USGS-understanding wells sampled for the Monterey Bay and Salinas Valley Basins study unit, California GAMA Priority Basin Project 50

Figure 17. Graph of well altitude at land surface, and total dissolved solid concentrations, in grid and understanding wells in the Santa Cruz, Monterey Bay, Salinas Valley, and Paso Robles study areas of the Monterey Bay and Salinas Valley Basins study unit, California GAMA Priority Basin Project 51

Figure 18. Graph of detection frequency and maximum relative-concentration of organic and special-interest constituents detected in USGS-grid wells in the Monterey Bay and Salinas Valley Basins study unit, California GAMA Priority Basin Project 52

Figures—Continued

Figure 19. Graph of detection frequency and relative-concentrations of selected organic and special-interest constituents in USGS-grid wells in the Monterey Bay and Salinas Valley Basins study unit, California GAMA Priority Basin Project, July–October 2005 53

Figure 20. Map showing land-use classifications and relative-concentrations of the herbicide simazine in U.S. Geological Survey (USGS)-grid wells (2002–2005), and for the period (July 17, 2002–July 18, 2005) from the California Department of Public Health (CDPH) database, Monterey Bay and Salinas Valley Basins study unit, California GAMA Priority Basin Project 55

Tables

Table 1. Number of wells sampled for the fast, intermediate, and slow sampling schedules, and number of constituents sampled in each constituent class, for the Monterey Bay and Salinas Valley Basins study unit, California GAMA Priority Basin Project, July–October 2005 14

Table 2. Inorganic constituents and associated benchmark information, and number of grid wells with U.S. Geological Survey-GAMA data and CDPH data, for each constituent, Monterey Bay and Salinas Valley Basins study unit, California GAMA Priority Basin Project 16

Table 3. Comparison of the number of compounds and median laboratory reporting levels or method detection limits by type of constituent for data reported in the California Department of Public Health (CDPH) database and for data collected by the U.S. Geological Survey (USGS) for the Monterey Bay and Salinas Valley Basins study unit, California GAMA Priority Basin Project, July–October 2005 17

Table 4. Aquifer-scale proportions from grid-based and spatially weighted approaches for constituents with high relative-concentrations during July 17, 2002–July 18, 2005 from the California Department of Public Health (CDPH) database, or with moderate or high relative-concentrations in samples collected from USGS-grid wells (July–October 2005), Monterey Bay and Salinas Valley Basins study unit, California GAMA Priority Basin Project 18

Table 5. Constituents in the California Department of Public Health (CDPH) database at high concentrations from April 24, 1974–July 17, 2002, Monterey Bay and Salinas Valley Basins study unit, California GAMA Priority Basin Project 20

Table 6. Results of non-parametric (Spearman's *rho* method) analysis of correlations in grid and understanding wells between selected potential explanatory factors, Monterey Bay and Salinas Valley Basins study unit, California GAMA Priority Basin Project 30

Table 7. Results of Wilcoxon rank-sum tests on grid-well data used to determine significant differences between selected water-quality constituents grouped by potential explanatory factor classifications, Monterey Bay and Salinas Valley Basins study unit, California GAMA Priority Basin Project 34

Tables—Continued

Table 8. Number of constituents analyzed, and, number detected, by the U.S. Geological Survey, with associated benchmarks in each constituent class, Monterey Bay and Salinas Valley Basins study unit, California GAMA Priority Basin Project, July–October 2005 37

Table 9. Aquifer-scale proportions for constituent classes, Monterey Bay and Salinas Valley Basins study unit, California GAMA Priority Basin Project 38

Table 10. Results of non-parametric (Spearman's method) analysis of correlations between selected water-quality constituents and potential explanatory factors, Monterey Bay and Salinas Valley Basins study unit, California GAMA Priority Basin Project 47

Conversion Factors, Datums, and Abbreviations and Acronyms

Conversion Factors

Inch/Foot/Mile to SI

Multiply	By	To obtain
Length		
inch (in.)	2.54	centimeter (cm)
inch (in.)	25.4	millimeter (mm)
foot (ft)	0.3048	meter (m)
mile (mi)	1.609	kilometer (km)
Area		
square foot (ft^2)	0.09290	square meter (m^2)
square mile (mi^2)	2.590	square kilometer (km^2)

Temperature in degrees Celsius (°C) may be converted to degrees Fahrenheit (°F) as follows:

$$°F=(1.8×°C)+32.$$

Temperature in degrees Fahrenheit (°F) may be converted to degrees Celsius (°C) as follows:

$$°C=(°F-32)/1.8.$$

Specific conductance is given in microsiemens per centimeter at 25 degrees Celsius (µS/cm at 25 °C).

Concentrations of chemical constituents in water are given either in milligrams per liter (mg/L) or micrograms per liter (µg/L) or micrograms per liter (µg/L). One milligram per liter is equivalent to 1 part per million (ppm); 1 microgram per liter is equivalent to 1 part per billion (ppb); 1 nanogram per liter (ng/L) is equivalent to 1 part per trillion (ppt); 1 per mil is equivalent to 1 part per thousand.

Conversion Factors, Datums, and Abbreviations and Acronyms—Continued

Datums

Vertical coordinate information is referenced to the North American Vertical Datum of 1988 (NAVD 88).

Horizontal coordinate information is referenced to the North American Datum of 1983 (NAD 83).

Abbreviations and Acronyms

AB	Assembly Bill (through the California State Assembly)
AL-US	U.S. Environmental Protection Agency action level
GAMA	Groundwater Ambient Monitoring and Assessment Program
HAL-US	U.S. Environmental Protection Agency lifetime health advisory level
HBSL	Health-based screening level
LRL	laboratory reporting level
LSD	land-surface datum
LT-MDL	long-term method detection level
MB	Monterey Bay study area of the Monterey Bay-Salinas Valley Basins GAMA study unit
MBFP	Monterey Bay study area flow-path well
MCL	maximum contaminant level
MCL-CA	California Department of Public Health maximum contaminant level
MCL-US	U.S. Environmental Protection Agency maximum contaminant level
MDL	method detection limit
MRL	minimum reporting level
MS	Monterey-Salinas GAMA Priority Basin Project study unit
NL-CA	California Department of Public Health notification level
pmc	percent modern carbon
PR	Paso Robles study area of the Monterey Bay-Salinas Valley Basins GAMA study unit
RSD5-US	U.S. Environmental Protection Agency risk-specific dose at a risk factor of 10^{-5}
SC	Santa Cruz study area of the Monterey Bay-Salinas Valley Basins GAMA study unit
SMCL	secondary maximum contaminant level
SMCL-CA	California Department of Public Health secondary maximum contaminant level
SMCL-US	U.S. Environmental Protection Agency secondary maximum contaminant level
SV	Salinas Valley study area of the Monterey Bay-Salinas Valley Basins GAMA study unit
TEAP	Terminal Electron Acceptor Processes
TT-US	treatment technique levels
TU	tritium unit
US	United States

Conversion Factors, Datums, and Abbreviations and Acronyms—Continued

Organizations

CDPH	California Department of Public Health (Department of Health Services prior to July 1, 2007)
CDPR	California Department of Pesticide Regulation
CDWR	California Department of Water Resources
LLNL	Lawrence Livermore National Laboratory
NAWQA	National Water-Quality Assessment Program (USGS)
NCDC	National Climatic Data Center
SWRCB	State Water Resources Control Board (California)
USEPA	U.S. Environmental Protection Agency
USGS	U.S. Geological Survey

Selected Chemical Names

MTBE	methyl *tert*-butyl ether
NDMA	*N*-nitrosodimethylamine
Nitrate-N	nitrate as nitrogen
Nitrite-N	nitrite as nitrogen
PCE	tetrachloroethene
TCE	trichloroethene
1,2,3-TCP	1,2,3-trichloropropane
TDS	total dissolved solids
THM	trihalomethane
VOC	volatile organic compound

Status and Understanding of Groundwater Quality in the Monterey Bay and Salinas Valley Basins, 2005: California GAMA Priority Basin Project

By Justin T. Kulongoski and Kenneth Belitz

Abstract

Groundwater quality in the approximately 1,000 square mile (2,590 km^2) Monterey Bay and Salinas Valley Basins (MS) study unit was investigated as part of the Priority Basin Project of the Groundwater Ambient Monitoring and Assessment (GAMA) Program. The study unit is located in central California in Monterey, Santa Cruz, and San Luis Obispo Counties. The GAMA Priority Basin Project is being conducted by the California State Water Resources Control Board in collaboration with the U.S. Geological Survey (USGS) and the Lawrence Livermore National Laboratory.

The GAMA MS study was designed to provide a spatially unbiased assessment of the quality of untreated (raw) groundwater in the primary aquifer systems (hereinafter referred to as primary aquifers). The assessment is based on water-quality and ancillary data collected in 2005 by the USGS from 97 wells and on water-quality data from the California Department of Public Health (CDPH) database. The primary aquifers were defined by the depth intervals of the wells listed in the CDPH database for the MS study unit. The quality of groundwater in the primary aquifers may be different from that in the shallower or deeper water-bearing zones; shallow groundwater may be more vulnerable to surficial contamination.

The first component of this study, the status of the current quality of the groundwater resource, was assessed by using data from samples analyzed for volatile organic compounds (VOC), pesticides, and naturally occurring inorganic constituents, such as major ions and trace elements. This *status assessment* is intended to characterize the quality of groundwater resources in the primary aquifers of the MS study unit, not the treated drinking water delivered to consumers by water purveyors.

Relative-concentrations (sample concentration divided by the health- or aesthetic-based benchmark concentration) were used for evaluating groundwater quality for those constituents that have Federal and (or) California regulatory or non-regulatory benchmarks for drinking-water quality. A relative-concentration greater than (>) 1.0 indicates a concentration greater than a benchmark, and less than or equal to ($\leq$) 1.0 indicates a concentration less than or equal to a benchmark. Relative-concentrations of organic and special interest constituents [perchlorate, *N*-nitrosodimethylamine (NDMA), and 1,2,3-trichloropropane (1,2,3-TCP)], were classified as "high" (relative-concentration > 1.0), "moderate" (0.1 < relative-concentration $\leq$ 1.0), or "low" (relative-concentration $\leq$ 0.1). Relative-concentrations of inorganic constituents were classified as "high" (relative-concentration > 1.0), "moderate" (0.5 < relative-concentration $\leq$ 1.0), or "low" (relative-concentration $\leq$ 0.5).

Aquifer-scale proportion was used as the primary metric in the *status assessment* for evaluating regional-scale groundwater quality. High aquifer-scale proportion was defined as the percentage of the area of the primary aquifers with a relative-concentration greater than 1.0 for a particular constituent or class of constituents; percentage is based on an areal rather than a volumetric basis. Moderate and low aquifer-scale proportions were defined as the percentage of the primary aquifers with moderate and low relative-concentrations, respectively. Two statistical approaches—grid-based and spatially weighted—were used to evaluate aquifer-scale proportions for individual constituents and classes of constituents. Grid-based and spatially-weighted estimates were comparable in the MS study unit (within 90-percent confidence intervals).

Inorganic constituents with human-health benchmarks were detected at high relative-concentrations in 14.5 percent of the primary aquifers, moderate in 35.5 percent, and low in 50.0 percent. High aquifer-scale proportion of inorganic constituents primarily reflected high aquifer-scale proportions of nitrate (7.9 percent), molybdenum (2.9 percent), arsenic (2.8 percent), boron (1.9 percent), and gross alpha-beta radioactivity (1.5 percent).

Relative-concentrations of organic constituents (one or more) were high in 0.2 percent, moderate in 6.6 percent, and low in 93.2 percent (not detected in 48.1 percent) of the primary aquifers. The high aquifer-scale proportion of organic constituents primarily reflected high aquifer-scale proportions of tetrachloroethene (0.1 percent) and methyl *tert*-butyl ether (0.1 percent). Relative-concentration for inorganic constituents with secondary maximum contaminant levels, manganese, total dissolved solids, iron, sulfate, and chloride were high in 18.6, 8.6, 7.1, 2.9, and 1.4 percent of the primary aquifers, respectively. Of the 205 organic and special-interest constituents analyzed, 32 constituents were detected. One organic constituent, the herbicide simazine, was frequently detected (in 10 percent or more of samples), but was detected at low relative-concentrations.

The second component of this study, the *understanding assessment*, identified the natural and human factors that affect groundwater quality by evaluating land use, physical characteristics of the wells, and geochemical conditions of the aquifer. Results from these evaluations were used to explain the occurrence and distribution of constituents in the study unit. The *understanding assessment* indicated that most wells that contained nitrate were classified as being in agricultural land-use areas, and depths to the top of perforations in most of the wells were less than 350 ft (76 m). High and moderate relative-concentrations of arsenic may be attributed to reductive dissolution of manganese or iron oxides, or to desorption or inhibition of arsenic sorption under alkaline conditions. Arsenic concentrations increased with increasing groundwater depth and residence time (age). Simazine was detected more often in groundwater from wells with surrounding land use classified as agricultural or urban, and with top of perforation depths less than 200 ft (61 m), than in groundwater from wells with natural land use or with deeper depths.

Tritium, helium-isotope, and carbon-14 data were used to classify the predominant age of groundwater samples into three categories: modern (water that has entered the aquifer since 1953), pre-modern (water that entered the aquifer prior to 1953 to tens of thousands of years ago), and mixed (mixtures of modern- and pre-modern-age waters). Arsenic concentrations were significantly greater in groundwater with pre-modern age classification than in groundwater with modern-age classification, suggesting that arsenic accumulates with groundwater residence time.

Introduction

To assess the quality of ambient groundwater in aquifers used for drinking-water supply and to establish a baseline groundwater-quality monitoring program, the State Water Resources Control Board (SWRCB), in collaboration with the U.S. Geological Survey (USGS) and Lawrence Livermore National Laboratory (LLNL), implemented the Groundwater Ambient Monitoring and Assessment (GAMA) Program (California Environmental Protection Agency, 2010, website at http://www.waterboards.ca.gov/gama/). The statewide GAMA program currently consists of three projects: the (1) GAMA Priority Basin Project, conducted by the USGS (U.S. Geological Survey, 2010, website at http://ca.water.usgs.gov/gama/); (2) the GAMA Domestic Well Project, conducted by the SWRCB; and (3) the GAMA Special Studies, conducted

by LLNL. On a statewide basis, the Priority Basin Project focused primarily on the deep portion of the groundwater resource, and the SWRCB Domestic Well Project generally focused on the shallow aquifer systems. The primary aquifers may be at less risk of contamination than the shallow wells, such as private domestic and environmental monitoring wells, which are closer to surficial sources of contamination. As a result, concentrations of constituents, such as volatile organic compounds (VOCs) and nitrate, in wells screened in the deep primary aquifers may be lower than concentrations of constituents in shallow wells (Kulongoski and others, 2010; Landon and others, 2010).

The SWRCB initiated the GAMA Program in 2000 in response to Legislative mandates (State of California, 1999, 2001a, Supplemental Report of the 1999 Budget Act 1999–00 Fiscal Year). The GAMA Priority Basin Project was initiated in response to the Groundwater Quality Monitoring Act of 2001 (State of California, 2001b, Sections 10780–10782.3 of the California Water Code, Assembly Bill 599) to assess and monitor the quality of groundwater in California. The GAMA Priority Basin Project is a comprehensive assessment of statewide groundwater quality, designed to help better understand and identify risks to groundwater resources and to increase the availability of information about groundwater quality to the public. For the Priority Basin Project, the USGS, in collaboration with the SWRCB, developed a monitoring plan to assess groundwater basins through direct sampling of groundwater and other statistically reliable sampling approaches (Belitz and others, 2003; California State Water Resources Control Board, 2003). Additional partners in the GAMA Priority Basin Project include the California Department of Public Health (CDPH), the California Department of Pesticide Regulation (CDPR), the California Department of Water Resources (CDWR), and local water agencies and well owners (Kulongoski and Belitz, 2004).

The range of hydrologic, geologic, and climatic conditions that exist in California must be considered in an assessment of groundwater quality. Belitz and others (2003) partitioned the State into 10 hydrogeologic provinces, each with distinctive hydrologic, geologic, and climatic characteristics (fig. 1). All these hydrogeologic provinces include groundwater basins and subbasins designated by the CDWR (California Department of Water Resources, 2003). Groundwater basins generally consist of relatively permeable, unconsolidated deposits of alluvial or volcanic origin. Eighty percent of California's approximately 16,000 public-supply wells are in designated groundwater basins. Groundwater basins and subbasins were prioritized for sampling on the basis of the number of public-supply wells, with secondary consideration given to municipal groundwater use, agricultural pumping, the number of historically leaking underground fuel tanks, and registered pesticide applications (Belitz and others, 2003). The 116 priority basins and additional areas outside defined groundwater basins were grouped into 35 study units, which include approximately 95 percent of public-supply wells in California.

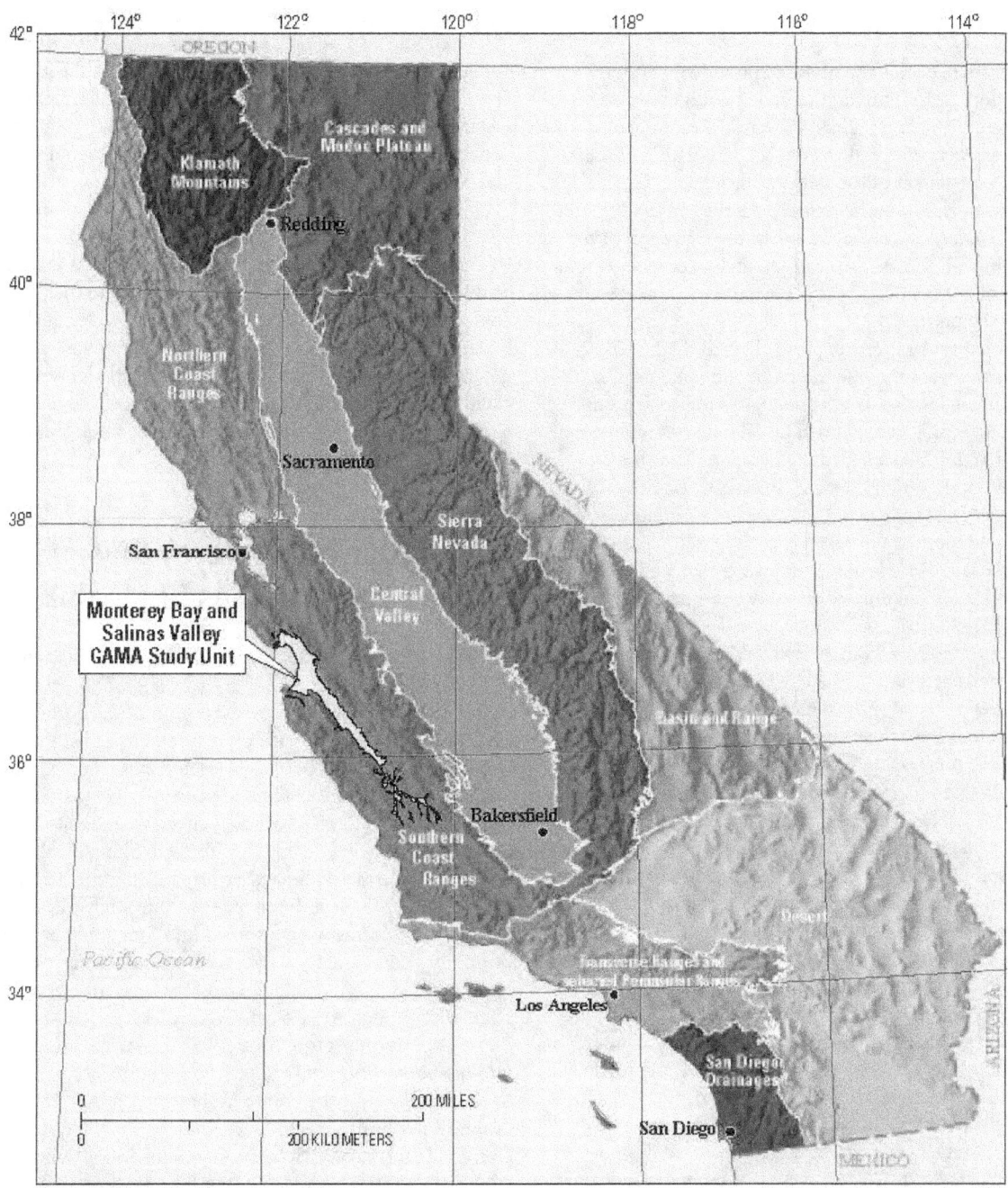

Base from U.S. Geological Survey National Elevation
Dataset, 2006, Albers Equal-Area Conic Projection

Provinces from Belitz and others, 2003.

Figure 1.   Monterey Bay and Salinas Valley Basins study unit, California GAMA Priority Basin Project, and the California hydrogeologic provinces.

Purpose and Scope

The purposes of this report are to provide a (1) study unit description: description of the hydrogeologic setting of the Monterey Bay and Salinas Valley Basins study unit (fig. 1), hereinafter referred to as the MS study unit, (2) *status assessment*: assessment of the status of the current (2005) quality of groundwater in the primary aquifers in the MS study unit, and (3) *understanding assessment*: identification of the natural and human factors affecting groundwater quality, and explanation of the relations between water quality and selected explanatory factors.

Water-quality data for samples collected by the USGS for the GAMA program in the MS study unit, and details of sample collection, analysis, and quality-assurance procedures for the MS study unit, are reported by Kulongoski and Belitz (2007). Utilizing those same data, this report describes methods used in designing the sampling network, identifying CDPH data for use in the *status assessment*, estimating aquifer-scale proportions of relative-concentrations, analyzing ancillary data sets, classifying groundwater age, and assessing the status and understanding of groundwater quality by statistical and graphical approaches.

The *status assessment* includes analyses of water-quality data for 91 wells selected by the USGS for spatial coverage of one well per grid cell (hereinafter referred to as USGS-grid wells) across the MS study unit. Most of these USGS-grid wells were public-supply wells, but 3 domestic and 11 irrigation wells with perforated-interval depths similar to the USGS-grid wells also were sampled. Samples were collected for analysis of anthropogenic constituents, such as VOCs and pesticides, and of naturally occurring inorganic constituents such as major ions and trace elements. Water-quality data from the California Department of Public Health (CDPH) database also were used to supplement data collected by the USGS for the GAMA program. The resulting set of water-quality data from USGS-grid wells and selected CDPH wells was considered to be representative of the primary aquifer systems (hereinafter referred to as primary aquifers) in the MS study unit; the primary aquifers are defined by the perforated-interval depths of the wells listed in the CDPH database for the MS study unit. GAMA *status assessment*s are designed to provide a statistically robust characterization of groundwater quality in the primary aquifers at the basin-scale (Belitz and others, 2003). The statistically robust design also allows basins to be compared and results to be synthesized regionally and statewide.

To provide context, the water-quality data discussed in this report were compared to California and Federal regulatory and non-regulatory benchmarks for treated drinking water. The assessments in this report are intended to characterize the quality of untreated groundwater resources in the primary aquifers within the study unit, not the treated drinking water delivered to consumers by water purveyors. The water delivered to consumers, after withdrawal from the ground, is typically treated, disinfected, and (or) blended with other waters to maintain acceptable water quality. Regulatory benchmarks apply to treated water that is delivered to the consumer, not to untreated groundwater.

The *understanding assessment* included data from 6 wells sampled by the USGS (hereinafter referred to as USGS-understanding wells) in addition to the 91 USGS-grid wells sampled for the *status assessment* to identify the natural and human factors affecting groundwater quality and to explain the relations between water quality and selected potential explanatory factors. Potential explanatory factors examined included land use, well depth, position of wells along the groundwater flowpath, indicators of groundwater age, and geochemical conditions.

Description of Monterey Bay and Salinas Valley Basins Study Unit

The MS study unit covers approximately 1,000 mi^2 (2,590 km^2) in Monterey, Santa Cruz, and San Luis Obispo Counties across the central coast region of California. The MS study unit lies within the Southern Coast Ranges hydrogeologic province (fig. 1) (Belitz and others, 2003) and includes eight groundwater basins (fig. 2): Santa Cruz Purisima Formation Highlands, Felton Area, Scotts Valley, Soquel Valley, West Santa Cruz Terrace, Salinas Valley, Pajaro Valley, and Carmel Valley (California Department of Water Resources, 2003). For the purpose of this study, these eight groundwater basins were grouped into four study areas based primarily on geography—the groundwater basins located near the town of Santa Cruz: the Santa Cruz Purisima Formation Highlands, Felton Area, Scotts Valley, Soquel Valley, and West Santa Cruz Terrace groundwater basins—were aggregated into the Santa Cruz study area. The groundwater basins east of Monterey Bay—the Pajaro Valley and Salinas Valley groundwater basins (including Langley, East Side Aquifer, Corral de Tierra Area, Seaside Area, and 180/400-Foot Aquifer subbasins)—were aggregated into the Monterey Bay study area. The Forebay Aquifer and Upper Valley Aquifer subbasins of the Salinas Valley groundwater basin were aggregated into the Salinas Valley study area. The Paso Robles Area subbasin of the Salinas Valley groundwater basin was established as the Paso Robles study area (fig. 2). As part of the Priority Basin Project, samples of untreated groundwater were collected from 97 wells in the MS study unit from July 18 to September 23, 2005 (Kulongoski and Belitz, 2007).

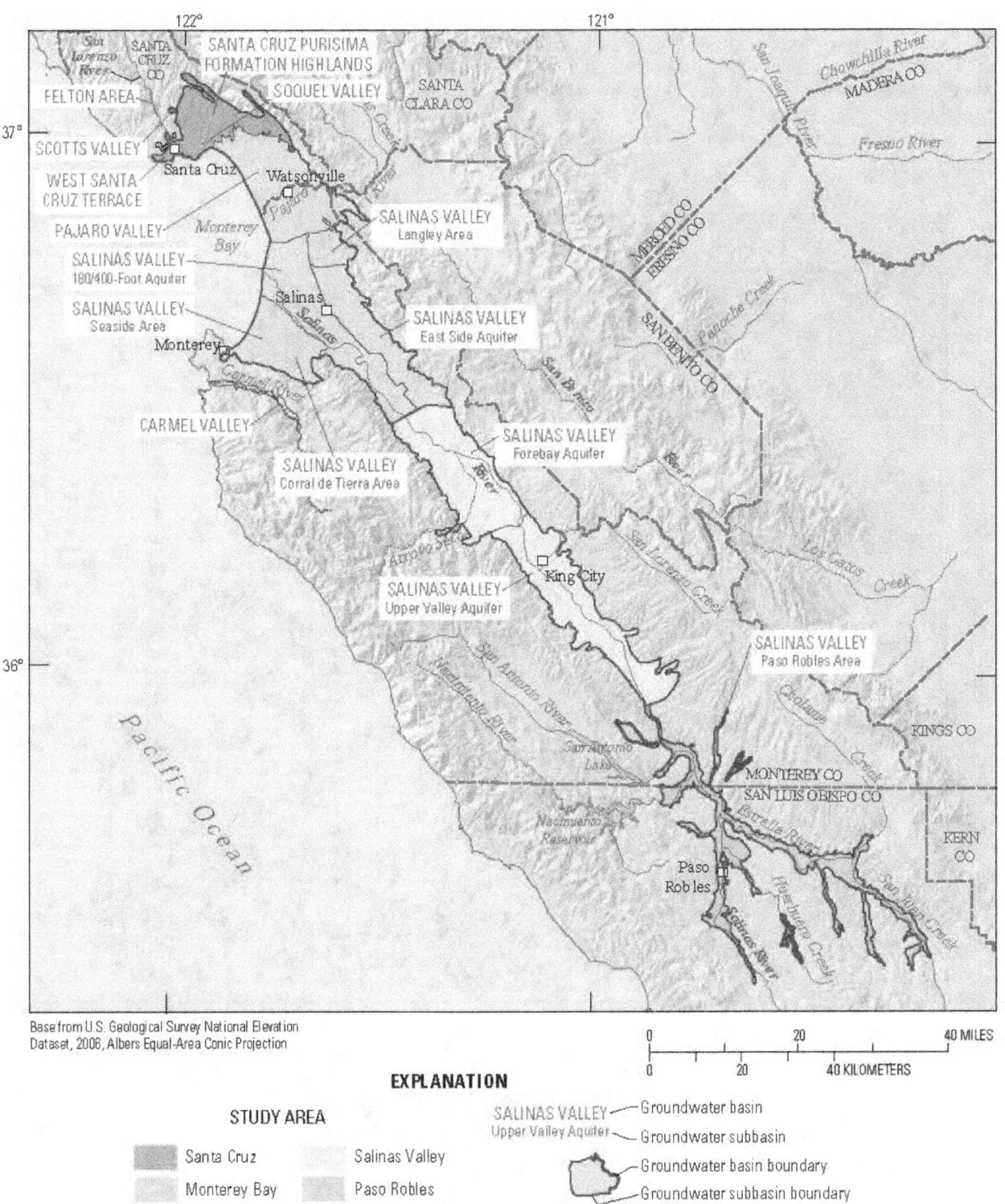

Base from U.S. Geological Survey National Elevation
Dataset, 2006, Albers Equal-Area Conic Projection

EXPLANATION

STUDY AREA

Santa Cruz   Salinas Valley

Monterey Bay   Paso Robles

SALINAS VALLEY — Groundwater basin
Upper Valley Aquifer — Groundwater subbasin

Groundwater basin boundary

Groundwater subbasin boundary

Figure 2. Geographic features and study areas of the Monterey Bay and Salinas Valley Basins study unit, California GAMA Priority Basin Project.

The Salinas Valley is the largest of the intermontane valleys of the Southern Coast Ranges and extends southeastward 120 mi (193 km) from Monterey Bay to Paso Robles (fig. 3). The Salinas Valley formed, in part, as a result of normal faulting along the King City (Rinconada-Reliz) Fault along the western margin of the valley from King City in the south to Monterey Bay in the north (fig. 2; 3) (California Department of Water Resources, 2003). Normal movement along the fault, valley-side down, resulted in the deposition of a westward thickening alluvial wedge (Showalter and others, 1983). The Salinas Valley is filled as much as 10,000 ft (3,048 m) on the east and as much as 15,000 ft (4,572 m) on

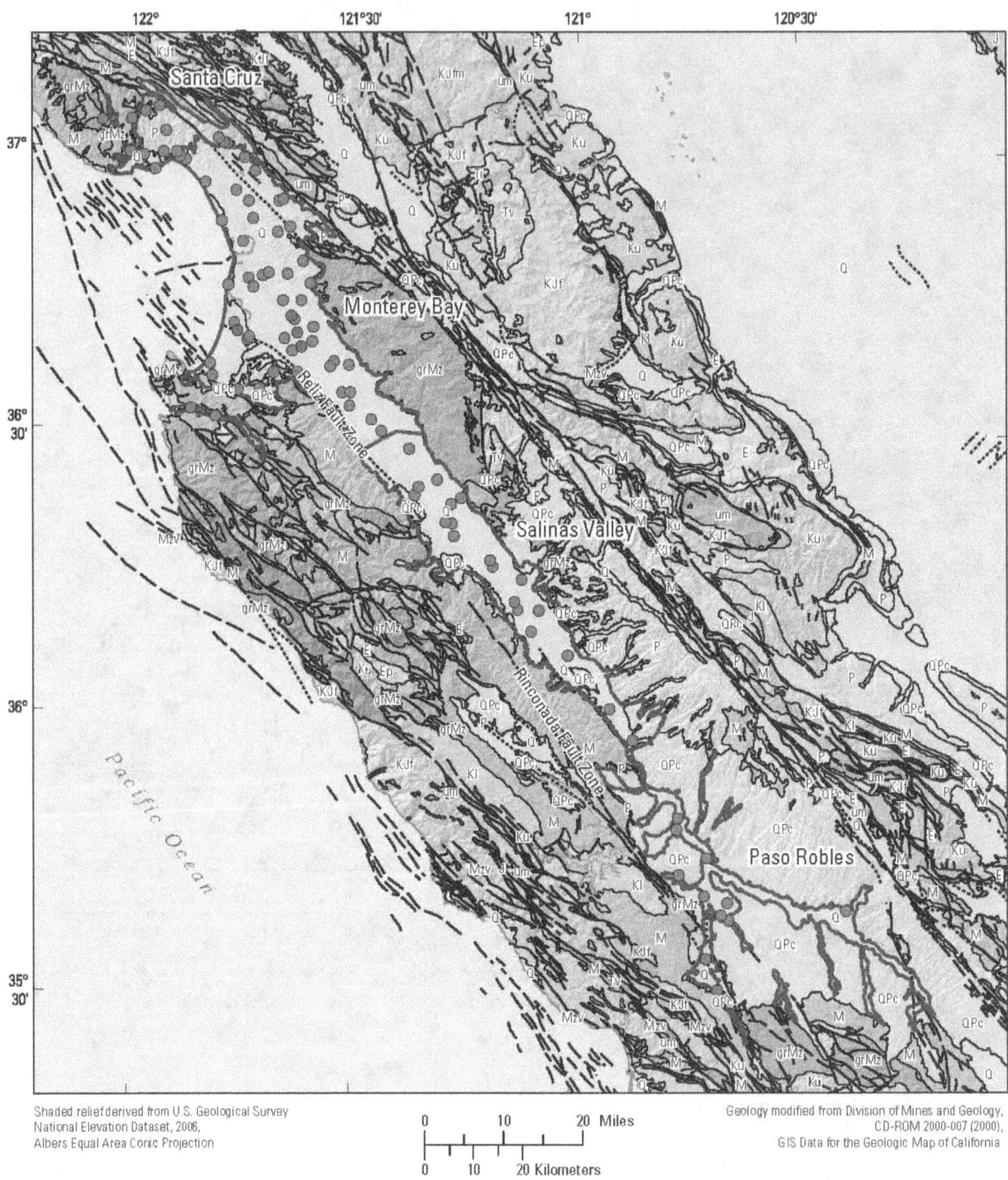

Figure 3.   Geologic formations and areal distribution of USGS grid and understanding wells sampled in the Monterey Bay and Salinas Valley Basins study unit, California GAMA Priority Basin Project.

the west with Tertiary and Quaternary marine and terrestrial sediments that include as much as 2,000 ft (609 m) of saturated alluvium (Showalter and others, 1983). Water-bearing units, which lie above mostly non-water-bearing and consolidated granitic basement, include the Miocene-age Monterey Formation, Pliocene-age Purisima Formation and

Pliocene- to Pleistocene-age Paso Robles Formation, and Pleistocene to Holocene alluvium (Hanson and others, 2002) (fig. 3). The primary aquifers that are the focus of the GAMA Primary Basin Project represent the water-bearing units that supply water for wells listed in the CDPH database.

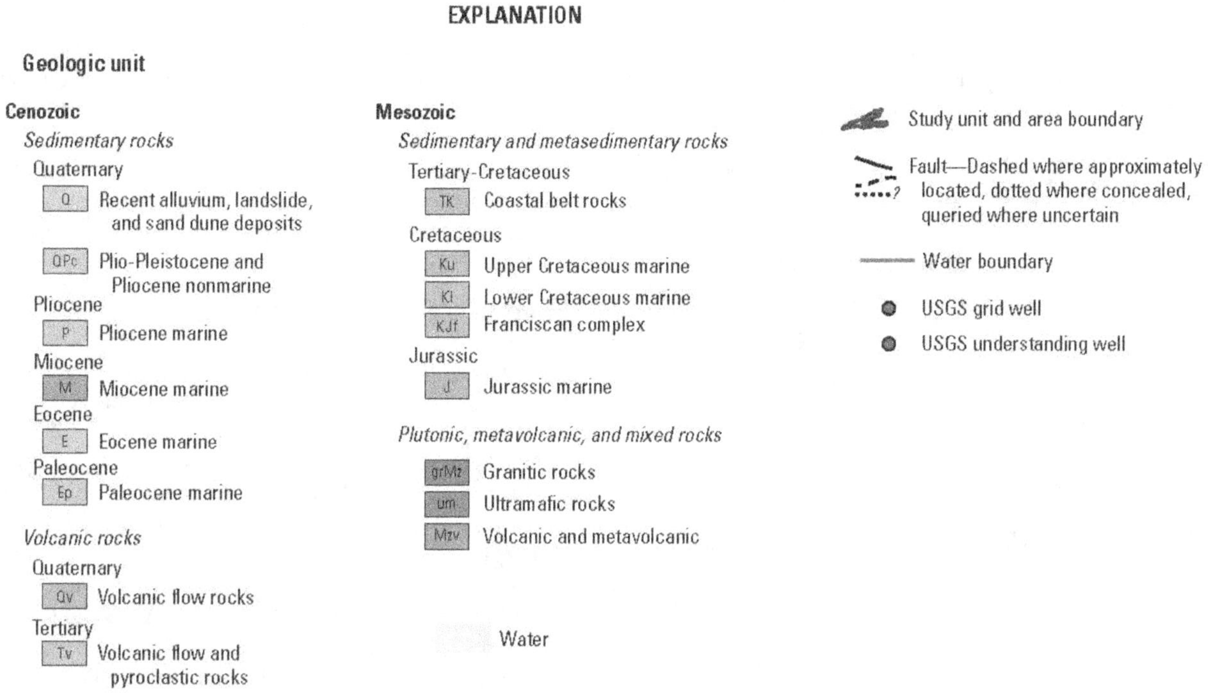

EXPLANATION

Geologic unit

Cenozoic

Sedimentary rocks

Quaternary

Q   Recent alluvium, landslide, and sand dune deposits

QPc   Plio-Pleistocene and Pliocene nonmarine

Pliocene

P   Pliocene marine

Miocene

M   Miocene marine

Eocene

E   Eocene marine

Paleocene

Ep   Paleocene marine

Volcanic rocks

Quaternary

Qv   Volcanic flow rocks

Tertiary

Tv   Volcanic flow and pyroclastic rocks

Mesozoic

Sedimentary and metasedimentary rocks

Tertiary-Cretaceous

TK   Coastal belt rocks

Cretaceous

Ku   Upper Cretaceous marine

Kl   Lower Cretaceous marine

KJf   Franciscan complex

Jurassic

J   Jurassic marine

Plutonic, metavolcanic, and mixed rocks

grMz   Granitic rocks

um   Ultramafic rocks

Mzv   Volcanic and metavolcanic

Water

Study unit and area boundary

Fault—Dashed where approximately located, dotted where concealed, queried where uncertain

Water boundary

USGS grid well

USGS understanding well

Figure 3.—Continued

Santa Cruz Study Area

The Santa Cruz (SC) study area, located in the northern part of the study unit (fig. 2), includes the Felton Area, Scotts Valley, Santa Cruz Purisima Formation Highlands, West Santa Cruz Terrace, and Soquel Valley groundwater basins (California Department of Water Resources, 2003). For the purposes of this study, these groundwater basins were grouped into the SC study area on the basis of the underlying Purisima Formation geology of the area; however, two wells near the town of Felton were sampled to represent the Felton groundwater basin, which is metamorphic terrain (fig. 3). The SC study area is bounded to the north, east, and west by the Santa Cruz Mountains, with altitudes as high as 2,900 ft (883 m), and is bounded to the south by Monterey Bay and the Pajaro Valley groundwater basin.

Mean annual precipitation at Santa Cruz is 31 in. (0.79 m) and mean annual temperature is 57°F (13.9°C), based on a 50-year record from the National Climatic Data Center (NCDC). The SC study area is drained by the San Lorenzo River and numerous creeks and their tributaries (fig. 2). Sources of groundwater recharge include percolation of rainfall, and river and stream infiltration.

In the northern part of the SC study area, the Santa Cruz Purisima Formation Highlands groundwater basin is defined by the geologic boundary of the Purisima Formation (fig. 3). The Upper-Pliocene Purisima Formation is the primary water-bearing unit and consists of poorly consolidated, silty to clean, very fine to medium-grained sandstone beds interbedded with siltstone. The formation ranges in thickness from 600 ft (183 m) in the north to 1,000 ft (305 m) in the south near Soquel (Muir, 1980).

The West Santa Cruz Terrace and Soquel Valley groundwater basins lie to the south of the Santa Cruz Purisima Formation Highlands groundwater basin. In the Soquel Valley groundwater basin, the water-bearing sediments consist of the Pliocene Purisima Formation, overlain by the Pleistocene Aromas Sand Formation and by Quaternary terrace deposits. The Purisima Formation and Quaternary terrace deposits have been incised locally by streams, and these channels have been filled with Quaternary alluvium (Muir, 1980). The Purisima Formation is a sequence of gray-to-blue, moderately consolidated, silty to clean, fine- to medium-grained sandstone containing siltstone and claystone interbeds (Greene, 1970). To the southeast, the Purisima Formation is overlain by hydraulically unconfined Aromas Sand Formation. The Aromas Sand Formation is brown to red, poorly consolidated, fine to coarse-grained sandstone containing lenses of silt and clay (California Department of Water Resources, 2003). The West Santa Cruz Terrace groundwater basin contains water-bearing sediments derived from the Purisima Formation, Quaternary terrace deposits, and alluvium along the San Lorenzo River and other streams (fig. 2). The Purisima Formation, the main water-bearing formation, is a thick sedimentary sequence with a fossiliferous marine rock base that grades to continental deposits in its upper portion. The

thin terrace deposits and alluvium are poorly cemented, moderately permeable gravel, sands, silts and silty clays, and yield only minor quantities of groundwater to wells (Greene, 1970).

The Scotts Valley and Felton Area groundwater basins are small alluvial valleys located in the Santa Cruz Mountains (figs. 2 and 4). The 2-mi² (5.2 km²) Felton Area groundwater basin and the 1.2-mi² (3.1 km²) Scotts Valley groundwater basin include the following formations from oldest to youngest: granitic basement, Tertiary Lompico Sandstone, Monterey Shale, Santa Margarita Sandstone, and Quaternary alluvium. The principal water-bearing formation is the unconfined Santa Margarita Sandstone, which is as much as 350 ft (107 m) thick. The underlying Lompico Sandstone also yields water, but to a lesser extent, and is as much as 600 ft (183 m) thick.

Monterey Bay Study Area

The Monterey Bay (MB) study area, as defined for the MS study unit, extends from east of Santa Cruz south along the Monterey Bay to the Forebay of the Salinas Valley. The MB study area covers approximately 450 mi² (1,166 km²) and includes most of the Quaternary sediment filled basins in this area (fig. 3), which include the Pajaro Valley and Carmel Valley groundwater basins, and the following subbasins of the Salinas Valley groundwater basin—180/400-Foot Aquifer, Eastside Aquifer, Seaside Area, Langley Area, and Corral de Tierra Area—as defined by the California Department of Water Resources (2003). For the purposes of this study, these basins and subbasins were grouped together in the MB study area because these basins contain similar Quaternary deposits.

Mean annual precipitation at Monterey is 20 in. (0.51 m), and mean annual temperature is 57°F (13.9°C), on the basis of a 50-year record from the NCDC. The MB study area is drained by the Salinas, Pajaro, and Carmel Rivers and their tributaries (fig. 2). Sources of groundwater recharge include percolation of precipitation, agricultural return flow, and river and stream runoff infiltration in the unconfined areas, but surficial recharge is restricted in the confined areas. In the confined areas, recharge is from underflow originating in upper valley areas, and groundwater flows north and west towards the discharge zones in the walls of the submarine canyon in Monterey Bay (Greene, 1970; Durbin and others, 1978).

The MB study area is bounded to the west by Monterey Bay and to the southwest by the Sierra de Salinas Mountains, which have altitudes as high as 4,470 ft (1,363 m) (fig. 4). The MB study area is bounded to the northeast by the Santa Cruz Mountains and to the southeast by the Gabilan Range, which have altitudes as high as 3,450 ft (1,052 m). The study area is bounded to the north by the surface expression of the geologic contact between Quaternary alluvium of the Pajaro Valley and marine sedimentary deposits of the Pliocene Purisima Formation (California Department of Water Resources, 2003).

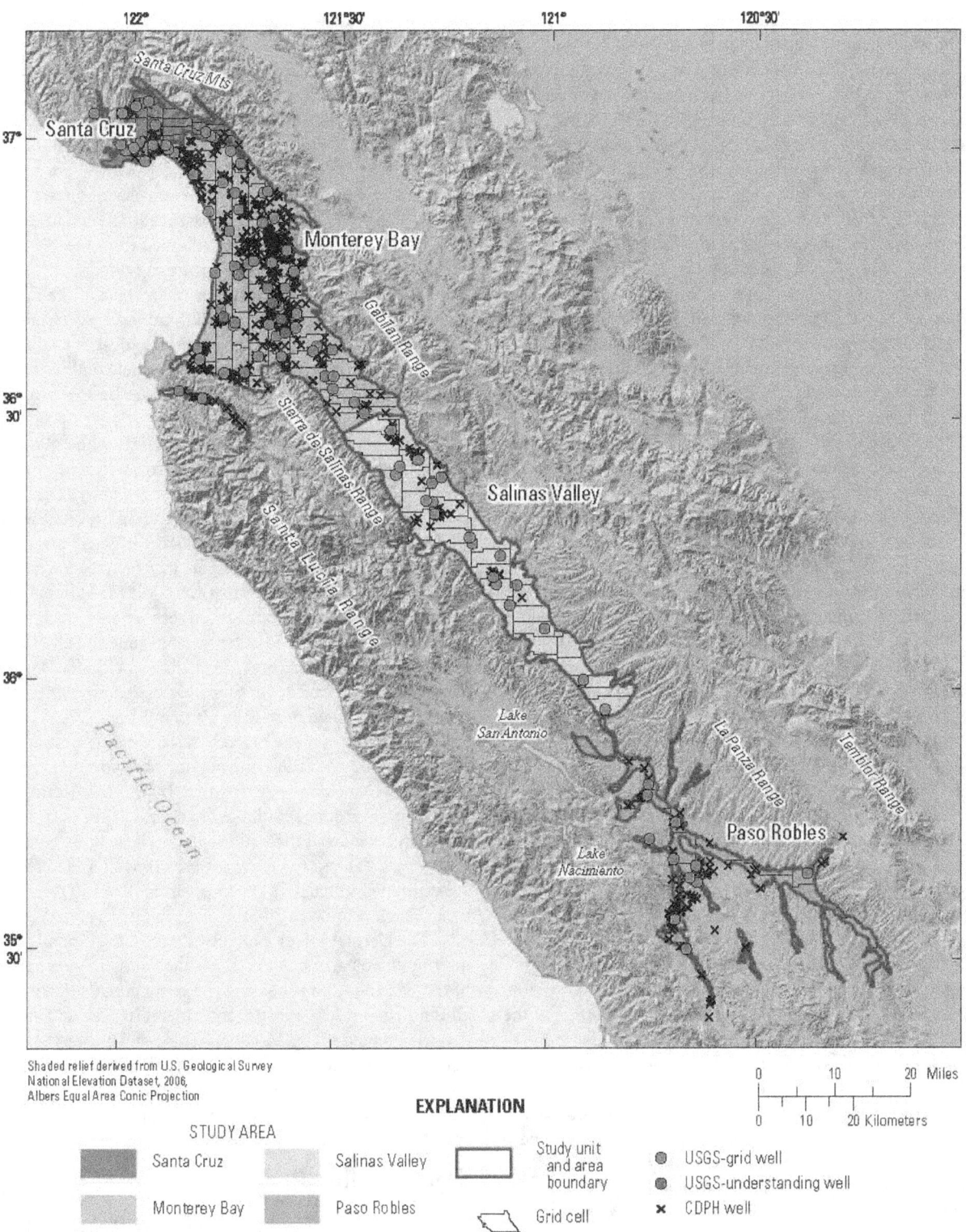

Shaded relief derived from U.S. Geological Survey
National Elevation Dataset, 2006,
Albers Equal Area Conic Projection

EXPLANATION

STUDY AREA

Santa Cruz

Monterey Bay

Salinas Valley

Paso Robles

Study unit
and area
boundary

Grid cell

⊙ USGS-grid well

● USGS-understanding well

× CDPH well

0   10   20 Miles

0   10   20 Kilometers

Figure 4. Locations of study area grid cells, U.S. Geological Survey (USGS) grid and understanding wells, and California Department of Public Health (CDPH) wells, Monterey Bay and Salinas Valley Basins study unit, California GAMA Priority Basin Project, July–October 2005.

The northern Pajaro Valley basin of the MB study area contains water-bearing geologic units that include, from oldest to youngest, the Purisima Formation, the Aromas Sand Formation, Terrace Deposits, Quaternary alluvium, and Dune Deposits (Johnson, and others, 1988). The Purisima Formation is mainly of marine origin, and contains a thick sequence of highly variable sediments ranging from shale beds near the base to continental deposits in the upper portion (Johnson and others, 1988). The sediments primarily are poorly consolidated, moderately permeable gravel, sands, silts, and silty clays (Johnson and others, 1988). The Aromas Sand Formation is composed of friable, quartzose, well-sorted brown to red sands that generally are medium-grained and weakly cemented with iron oxide (Johnson and others, 1988). This unit ranges in thickness from 100 ft (31 m) inland near the foothills, to nearly 900 ft (274 m) offshore near the mouth of the Pajaro River (Allen, 1946). The Aromas Sand, considered the primary water-bearing unit of the basin, consists of upper eolian and lower fluvial sand units that are separated by confining layers of interbedded clays and silty clay (Johnson and others, 1988). The Terrace Deposits consist of unconsolidated gravel, sand, silt, and clay overlain by alluvium. The alluvium is composed of Pleistocene terrace materials that are overlain by Holocene alluvium, consisting of sand, gravel, and clay deposited by the Pajaro River, and dune sands, with an average thickness of 50 to 300 ft (15 to 91 m). A 400-ft (122 m) deep, inland-projecting buried paleodrainage of the Salinas River acts as the southern subbasin boundary and restricts flow into the 180/400-Foot Aquifer subbasin.

South of the Pajaro Valley basin lay the 180/400-Foot Aquifer and Langley Area subbasins (fig. 2). The 24-mi² (62 km²) Langley Area subbasin is a series of low hills composed of the following formations, from oldest to youngest, the Pliocene to Pleistocene Paso Robles Formation, the Pleistocene Aromas Sands, Quaternary terrace deposit, Holocene alluvium, and sand dunes (California Department of Water Resources, 1977). Outcrops of the Aromas Sands compose most of the subbasin, but exposures of Quaternary terrace deposits and Holocene alluvium along creeks form a small portion of the southeastern subbasin. The lower portion of the Aromas Sands interfingers with the upper portion of the Paso Robles Formation to form the 400-Foot Aquifer to the west in the Salinas Valley 180/400-Foot Aquifer subbasin.

The 180/400-Foot Aquifer subbasin includes three water-bearing units, the 180-Foot, the 400-Foot, and the 900-Foot Aquifers, named for the average depths of each aquifer. The confined 180-Foot Aquifer occurs only in this subbasin, as its confining blue clay layer thins and disappears east of the subbasin. The 180-Foot Aquifer consists of interconnected sand, gravel, and clay lenses, and ranges in thickness from 50 ft (15 m) near Salinas, to 150 ft (46 m) near Monterey Bay (Durbin and others, 1978). The 180-Foot Aquifer is separated from the 400-Foot Aquifer by a zone of lesser aquifers and confining units that range in thickness from 10 to 70 ft (3 to 21 m). The 400-Foot Aquifer consists of sands, gravels, and clay lenses, with an average thickness of 200 ft (64 m) (Durbin and others, 1978). The upper portion of the aquifer may be correlative with the Aromas Sand and the lower portion with the upper part of the Paso Robles Formation (Montgomery-Watson Consulting Engineers, 1994). The 900-Foot Aquifer, present in the lower Salinas Valley, consists of alternating layers of sand, gravels and clays with a total thickness as much as 900 ft, (274 m) thick and is separated from the 400-Foot Aquifer by a blue marine clay -confining unit.

The 180/400-Foot Aquifer is to the west of the Eastside Aquifer subbasin. This 90-mi² (233 km²) subbasin includes the same water-bearing units as the 180/400-Foot Aquifer subbasin. However, the blue clay layer that confines the 180-Foot Aquifer does not extend into the Eastside Aquifer subbasin.

The 180/400-Foot Aquifer subbasin is to the north of the Seaside Area and Corral de Tierra Area subbasins. These subbasins include the following water-bearing units, from oldest to youngest: the Miocene and Pliocene Santa Margarita Formation, the Pliocene Paso Robles Formation, the Pleistocene Aromas Formation, and Pleistocene and Holocene age alluvial deposits (Muir, 1982). Although the aggregate maximum thickness of these units is greater than 1,000 ft (335 m), surface outcrops are limited to alluvial sand and terrace deposits (Muir, 1982). The Santa Margarita Formation has a maximum thickness of 225 ft (69 m), and is poorly consolidated marine sandstone (Muir, 1982). The Paso Robles Formation is the primary water-bearing unit in the area and consists of sand, gravel, and clay interbedded with some minor calcareous beds (Muir, 1982). The Aromas Sand Formation is grouped with the dune sand deposits within this subbasin because of their similarities. These units consist of relatively clean red to yellowish-brown, well-sorted sand and are estimated to range in thickness from 30 to 50 ft (9 to 15 m) near the coast to up to 200 ft (61 m) inland (Muir, 1982).

The Carmel Valley groundwater basin is a small intermontane basin that lies along the Carmel River south of the Seaside Area subbasin. The basin contains younger alluvium and river deposits, and older alluvium and terrace deposits, underlain by Monterey Shale and Tertiary sandstone units. The younger alluvium comprises the main water-bearing units and consists of boulders, gravel, sand, silt, and clay, with a thickness between 30 and 180 ft (Kapple and others, 1984).

Salinas Valley Study Area

The Salinas Valley (SV) study area (fig. 2) includes the following groundwater subbasins of the Salinas Valley basin: the Forebay Aquifer and the Upper Valley Aquifer, as defined by the California Department of Water Resources (2003). For the purposes of this study, these subbasins were combined into the SV study area based on similar geology of the upper and central Salinas Valley. The northern boundary of the SV study area is shared with the 180/400-Foot Aquifer and Eastside Aquifer subbasins. The SV study area is bounded to the west by the Sierra de Salinas and Santa Lucia Ranges, with altitudes as high as 4,850 ft (1,478 m), and to the east by the Gabilan Range (fig. 4). The southern boundary of the SV study area, at the constriction of the Salinas Valley where Sargent Creek joins the Salinas River, is shared with the Paso Robles Area subbasin and separates the upper and lower Salinas River drainage basins.

Mean annual precipitation at Salinas is 15 in. (0.38 m) and mean annual temperature is 58°F (14.4°C), based on a 50-year record from the NCDC. The SV study area is drained by the Salinas River and its tributaries. Sources of groundwater recharge include river and stream runoff infiltration and applied irrigation water.

The SV study area covers approximately 300 mi^2 (777 km^2) of the central Salinas Valley. The main water-bearing units of this subbasin are unconsolidated to semi-consolidated and interbedded gravel, sand and silt, alluvial-fan, and river deposits (Durbin and others, 1978). These deposits form the 180-Foot and 400-Foot Aquifers that were mentioned previously in the MB study area description. The northern boundary of the SV study area marks the southern boundary of the confining conditions for the 180-Foot Aquifer, and just south of Arroyo Seco in the center of the SV study area (the southern boundary of the Forebay Aquifer subbasin) marks the southern boundary of the confining conditions above the 400-Foot Aquifer. In the Forebay Aquifer subbasin, groundwater is found in the lenses of sand and gravel that are interbedded with massive units of finer grained material (Durbin and others, 1978). In the northern Forebay Aquifer subbasin, the unconfined 180-Foot Aquifer ranges in thickness from 50 to 150 ft (15 to 46 m), with an average thickness of 100 ft (30 m), and is separated from the 400-Foot Aquifer by a zone of discontinuous sands and blue clays called the 180/400-Foot confining unit. The aquiclude ranges in thickness from 10 to 70 ft (3 to 21 m) above the 400-Foot Aquifer, which has an average thickness of 200 ft (61 m) (Durbin and others, 1978). To the south, the Upper Aquifer subbasin, a lateral equivalent to the 180/400-Foot Aquifers, includes unconsolidated to semi-consolidated and interbedded gravel, sand, and silt of the Paso Robles Formation alluvial fan and river deposits, but the 400-Foot confining unit is absent in this portion of the valley.

An additional deep aquifer consisting of alternating layers of sand-gravel mixtures and clays, the 900-Foot Aquifer, is present in the Forebay Aquifer subbasin of the Salinas Valley, but does not extend into the Upper Valley Aquifer subbasin because of the southward shallowing of the basement complex (Durbin and others, 1978).

Paso Robles Study Area

The Paso Robles (PR) study area (fig. 2) lies within the Paso Robles Area subbasin of the Salinas Valley groundwater basin, as defined by the California Department of Water Resources (2003). For the purposes of this study, the Quaternary alluvium that fills the valleys in this subbasin is designated as the PR study area (fig. 2), which excludes the higher altitude Quaternary-Pleistocene deposits. The PR study area is bounded to the east by the Temblor Range, to the south by the La Panza Range, to the west by the Santa Lucia Range (fig. 4), and to the north by the Upper Salinas Valley Aquifer subbasin (California Department of Water Resources, 2003).

Mean annual precipitation at Paso Robles is 13 in. (0.33 m) and mean annual temperature is 60°F (15.6°C), based on a 50-year record from the NCDC. Sources of groundwater recharge include infiltration of precipitation, return flow from irrigation, and seepage from rivers and streams.

The PR study area covers approximately 300 mi^2 (777 km^2) of valley sediments in the low-lying areas along the San Antonio and Nacimiento Rivers in the west, the Salinas River and Huerhuero Creek in the south, the Estrella River in the center, and the San Juan Creek to the southeast (fig. 2). These rivers and their tributaries drain the PR study area. Water-bearing formations in this study area include the Quaternary alluvium, which consists of unconsolidated, fine- to coarse-grained sand with pebbles and boulders as much as 130 ft (39.6 m) thick near the Salinas River (California Department of Water Resources, 1999).

Hydrogeologic Setting

The climate in the MS study unit is characterized by warm, dry summers and cool, moist winters. At the National Climate Data Center (NCDC) station in Monterey, on the basis of a 50-year record, the average annual temperature is 57°F (13.9°C), and the average annual precipitation is 20 in. (0.51m), occurring as rain during the winter and early spring. However, the distribution of precipitation across the area is dependent on the topography and the prevailing winds, with an increase in precipitation concomitant to an increase in altitude. Precipitation also decreases with latitude from north to south in the MS study unit. Fifty-year climate records from NCDC stations from Santa Cruz to Paso Robles show that the mean annual precipitation decreases from 31 in. (0.79 m) in Santa Cruz in the north to 13 in. (0.33 m) in Paso Robles in the south.

The MS study unit groundwater basins are drained by several rivers and their principal tributaries, including the Salinas Valley drained by the Salinas River; the Pajaro Valley drained by the Pajaro River; the Santa Cruz area drained by the San Lorenzo River; and the Carmel Valley drained by the Carmel River (fig. 2).

Sources of groundwater recharge include percolation of precipitation, river and stream infiltration, and agricultural irrigation and return flow. The relative contributions of these inputs also are dependent on the hydrogeologic setting of each area.

In the study areas, the directions of groundwater flow generally follow the topography of the basins, from high altitudes towards the drainages, and down valleys towards the Monterey Bay and Pacific Ocean. Water resources for public drinking-water supply, and irrigation, include surface water from Lake San Antonio, Lake Nacimiento, Pinto Lake, and local public-supply wells. The primary aquifer targeted by this study includes groundwater-bearing zones in which public-supply wells (CDPH database) are completed. These wells range in depth from 69 to 1,950 ft (21 to 594 m), depending on well location and depth of the alluvium. Groundwater in the alluvium moves under a natural hydraulic gradient that conforms in a general way to the surface topography. Groundwater movement generally is from the southern part of the Salinas Valley northward towards the Monterey Bay.

Methods

The *status assessment* provides a spatially unbiased assessment of groundwater quality in the primary aquifers and the *understanding assessment* was designed to evaluate the natural and human factors that affect groundwater quality of the MS study unit. This section describes the methods used for: (1) defining groundwater quality, (2) assembling the datasets used for the *status assessment*, (3) determining which constituents warrant assessment, (4) calculating aquifer-scale proportions, and (5) providing statistical analyses for the *understanding assessment*. Methods used for compilation of data on potential explanatory factors are described in appendix A.

The primary metric for defining groundwater quality is *relative-concentration*, which references concentrations of constituents measured in groundwater to regulatory and non-regulatory benchmarks used to evaluate drinking-water quality. A subset of constituents was selected for additional evaluation in the assessment on the basis of objective criteria by use of these relative-concentrations. Groundwater-quality data collected by the U.S. Geological Survey for the GAMA Priority Basin Project (USGS–GAMA) and data compiled in the CDPH database are used in the *status assessment*. Two statistical methods based on spatially unbiased equal-area grids are used to calculate aquifer-scale proportions of low, moderate, or high relative-concentrations: (1) the "grid-based" method uses one value per grid cell to represent groundwater quality (Belitz and others, 2010), and (2) the "spatially weighted" method uses many values per grid cell.

The CDPH database contains historical records from more than 27,000 wells, necessitating targeted retrievals to effectively access relevant water-quality data. For example, for the area representing the MS study unit, the historical CDPH database contains more than 502,000 records from 850 wells. The CDPH data were used in three ways in the *status assessment*: (1) to fill in gaps in the USGS data for the grid-based calculations of aquifer-scale proportions, (2) to select constituents for additional evaluation in the assessment, and (3) to provide the majority of the data used in the spatially weighted calculations of aquifer-scale proportions.

Relative-Concentrations and Water-Quality Benchmarks

Concentrations of constituents are presented as relative-concentrations in the *status assessment*:

$$\text{Relative concentration} = \frac{\text{Sample concentration}}{\text{Benchmark concentration}}.$$

Relative-concentrations were used to provide context for the measured concentrations in the sample. Relative-concentrations less than 1 (<1.0) indicate a sample concentration less than the benchmark, and relative-concentrations greater than 1 (>1.0) indicate a sample concentration greater than the benchmark. The use of relative-concentrations also permits comparison on a single scale of constituents present at a wide range of concentrations.

Toccalino and others (2004), Toccalino and Norman (2006), and Rowe and others (2007) previously used the ratio of measured sample concentration to the benchmark concentration [either maximum contaminant levels (MCLs) or Health-Based Screening Levels (HBSL)] and defined this ratio as the Benchmark Quotient. Relative-concentrations used in this report are equivalent to the Benchmark Quotient reported by Toccalino and others (2004) for constituents with MCLs. However, HBSLs were not used in this report because HBSLs are not currently used as benchmarks by California drinking-water regulatory agencies. Relative-concentrations can only be computed for constituents with water-quality benchmarks; therefore, constituents without water-quality benchmarks are not included in the *status assessment*.

Regulatory and non-regulatory benchmarks apply to treated water that is served to the consumer, not to untreated groundwater. However, to provide some context for the results, concentrations of constituents measured in the untreated groundwater were compared to benchmarks established by the U.S. Environmental Protection Agency (USEPA) and CDPH (U.S. Environmental Protection Agency, 2006; California Department of Health Services, 2008a, 2008b). The benchmarks used for each constituent were selected in the following order of priority:

1. Regulatory, health-based CDPH and USEPA maximum contaminant levels (MCL-CA and MCL-US), action levels (AL-US), and treatment technique levels (TT-US).

2. Non-regulatory CDPH and USEPA secondary maximum contaminant levels (SMCL-CA and SMCL-US). For constituents with both recommended and upper SMCL-CA levels, the values for the upper levels were used.

3. Non-regulatory, health-based CDPH notification levels (NL-CA), USEPA lifetime health-advisory levels (HAL-US) and USEPA risk-specific doses for 1:100,000 (RSD5-US).

For constituents with multiple types of benchmarks, this hierarchy may not result in selection of the benchmark with the lowest concentration. Additional information on the types of benchmarks and listings of the benchmarks for all constituents analyzed is provided by Kulongoski and Belitz (2007).

For ease of discussion, relative-concentrations of constituents were classified into low, moderate, and high categories:

Category	Relative-concentrations for organic and special interest constituents	Relative-concentrations for inorganic constituents
High	> 1	> 1
Moderate	> 0.1 and ≤ 1	> 0.5 and ≤ 1
Low	≤ 0.1	≤ 0.5

For organic and special-interest constituents, a relative-concentration of 0.1 was used as a threshold to distinguish between low and moderate relative-concentrations for consistency with other studies and reporting requirements (U.S. Environmental Protection Agency, 1998; Toccalino and others, 2004). For inorganic constituents, a relative-concentration of 0.5 was used as a threshold to distinguish between low and moderate relative-concentrations. A larger threshold value was used because in the MS study unit, and elsewhere in California (Kulongoski and others, 2010), the naturally occurring inorganic constituents tend to be more prevalent than organic constituents in groundwater. Although more complex classifications could be devised based on the properties and sources of individual constituents, use of a single moderate/low threshold value for each of the two major groups of constituents provided a consistent, objective criteria for distinguishing constituents at moderate rather than low concentrations.

Datasets for Status Assessment

U.S. Geological Survey Grid Wells

The primary data used for the grid-based calculations of aquifer-scale proportions of relative-concentrations were data from wells sampled by USGS-GAMA. Detailed descriptions of the methods used to identify wells for sampling are given in Kulongoski and Belitz (2007). Briefly, each study area was divided into 10-mi^2 (~25 km^2) equal-area grid cells, and in each cell, one well was randomly selected to represent the cell (fig. 4) (Scott, 1990). Wells were selected from the population of wells in statewide databases maintained by the CDPH and the USGS. The MS study unit contained a total of 116 grid cells, and the USGS sampled wells in 91 of those cells (USGS-grid wells). Of the 91 USGS-grid wells, 77 were listed in the CDPH database; the other 10 were irrigation or domestic wells perforated at depths similar to the depths of CDPH wells in their respective cells, and 4 irrigation wells did not have well construction data available. USGS-grid wells were named with an alphanumeric GAMA ID consisting of a prefix identifying the study area and a number indicating the order of sample collection (fig. B1; table A1). The following prefixes were used to identify the study area: SC, Santa Cruz study area, MB, Monterey Bay study area, SV, Salinas Valley study area, and PR, Paso Robles study area.

Samples collected from USGS-grid wells were analyzed for 216 to 284 constituents (table 1). Water-quality indicators (field parameters), volatile organic compounds, pesticides, noble gases, and selected isotopes used as hydrologic tracers were analyzed in samples from all USGS wells. Major and minor ions, trace elements, nutrients, and redox species, radiochemical constituents, carbon isotopes, N-nitrosodimethylamine (NDMA), perchlorate, and 1,2,3-trichloropropane (1,2,3-TCP) were analyzed in samples from 31 wells. The collection, analysis, and quality-control data for the analytes listed in table 1 are described by Kulongoski and Belitz (2007). However, further quality assurance and quality controls (QA/QC) were applied to the data. Data for constituents detected in the field blank samples were screened for concentrations that were less than a concentration equal to the constituent's highest blank sample detection plus one-half of the constituent's laboratory reporting level (added to accommodate uncertainty in the laboratory analyses); results that were less than this screening level were considered to be nondetections for the purposes of this study.

Table 1. Number of wells sampled for the fast, intermediate, and slow sampling schedules, and number of constituents sampled in each constituent class, for the Monterey Bay and Salinas Valley Basins study unit, California GAMA Priority Basin Project, July–October 2005.

[1,2,3-TCP, 1,2,3-trichloropropane; NDMA, *N*-nitrosodimethylamine]

	Sampling schedule		
	Fast	Intermediate	Slow
Well summary	**Number of wells**		
Total number of wells	63	3	31
Number of grid wells sampled	62	0	29
Number of understanding wells sampled	1	3	2
Constituent class	**Number of constituents**		
Water-quality indicators (field parameters)			
Specific conductance and temperature	2	2	2
Dissolved oxygen and pH		2	2
Field alkalinity, bicarbonate, and carbonate		3	3
Turbidity			1
Organic constituents			
Volatile organic compounds (VOCs) and gasoline additives [1]	88	88	88
Pesticides and pesticide degradates	61	61	61
Polar pesticides and degradates	53	53	53
Dissolved organic carbon		1	1
Constituent of special interest			
Perchlorate, NDMA, and low-level 1,2,3-TCP [2]		3	3
Inorganic constituents			
Major and minor ions, silica, total dissolved solids (TDS), and trace elements		36	36
Nutrients		5	5
Arsenic and iron species		4	4
Chromium species	2	2	2
Isotopes			
Stable isotopes of hydrogen and oxygen	2	2	2
Carbon-13 and carbon-14			2
Radioactivity and dissolved gases			
Tritium [3]	1	1	1
Noble gases and tritium [4]	7	7	7
Radon and radium isotopes			3
Gross alpha and beta radioactivity		4	4
Microbial constituents			
Total coliforms, colifage (somatic and F-specific), *E. coli*			4
Total	216	274	284

[1] Includes nine constituents classified as fumigants or fumigant synthesis by-products.

[2] 1,2,3-TCP was analyzed as a constituent of special interest with a method reporting level of 0.005 µg/L (microgram per liter), and also on the U.S. Geological Survey VOC schedule 2020, which has a laboratory reporting level of 0.12 µg/L.

[3] Analyzed at U.S. Geological Survey Tritium Laboratory, Menlo Park, California.

[4] Analyzed at Lawrence Livermore National Laboratory, Livermore, California.

California Department of Public Health Grid Wells

The four study areas were divided into 116 grid cells: of these, 25 cells did not have a USGS-grid well (fig. B1–B2), and 62 cells had a USGS-grid well but no USGS data for major ions, trace elements, nutrients, and radiochemical constituents. The CDPH database was queried to provide these missing inorganic and radiochemical data. CDPH wells with data for the most recent 3 years available at the time of sampling (July 17, 2002–July 18, 2005) were considered. If a well had more than one analysis for a constituent in the 3-year interval, then the most recent data were selected.

The decision tree used to identify suitable data from CDPH wells is described in appendix B. Briefly, the first choice was to use CDPH data from the same well sampled by the USGS (USGS-grid well). In this case, "DG" was added to the well's GAMA ID to signify that it was a well sampled by the USGS that also used CDPH data (fig. B3–B4; table A1). If the DG well did not have all the needed data, then a second well in the cell was randomly selected from the subset of CDPH wells with data and a new identification with "DPH" and a new number was assigned to that well (fig. B3–B4; table A1). The combination of the USGS-grid wells and the DG- and DPH-CDPH grid wells produced a grid-well network covering 94 of the 116 grid cells in the MS study unit (table A1). No accessible wells or necessary data were available for the remaining 22 cells.

The CDPH database generally did not contain data for all missing inorganic constituents at every CDPH grid well; therefore, the number of wells used for the grid-based assessment differed for various inorganic constituents (table 2). Although other organizations also collect water-quality data, the CDPH data is the only statewide database of groundwater-chemistry data available for comprehensive analysis.

CDPH data were not used to supplement USGS-grid well data for VOCs, pesticides, or perchlorate for the *status assessment*. A larger number of VOCs and pesticide compounds are analyzed for the USGS-GAMA program than are available from the CDPH database. USGS-GAMA collected data for 88 VOCs plus 114 pesticides and pesticide degradates at each of the 97 wells sampled by the USGS in the MS study unit (table 1). In addition, method detection limits for USGS-GAMA analyses typically were one to two orders of magnitude lower than the reporting levels for analyses compiled by CDPH (table 3).

Additional Data Used for Spatially Weighted Calculation

The spatially weighted calculations of aquifer-scale proportions of relative-concentrations for the MS study unit used data from the USGS-grid wells, from additional wells sampled by USGS-GAMA, and from all wells in the CDPH database having water-quality data during the 3-year interval July 17, 2002–July 18, 2005. For wells with both USGS and CDPH data, only the USGS data were used.

Six additional wells were sampled by the USGS to increase the sampling density in the MB study area to better understand specific groundwater-quality issues (figs. B1–B2). These "USGS-understanding" wells were numbered with prefixes modified from those used for the USGS-grid wells (for example, MBFP01-03- and MBMW01-03) (figs. B1–B2; table A1).

Table 2. Inorganic constituents and associated benchmark information, and number of grid wells with U.S. Geological Survey-GAMA data and CDPH data, for each constituent, Monterey Bay and Salinas Valley Basins study unit, California GAMA Priority Basin Project.

[USGS, U.S. Geological Survey; CDPH, California Department of Public Health; HAL-US, USEPA lifetime health advisory level; MCL-US, USEPA maximum contaminant level; MCL-CA, CDPH maximum contaminant level; NL-CA, CDPH notification level; AL-US, USEPA action level; SMCL-CA, CDPH secondary maximum contaminant level; SMCL-US, USEPA secondary maximum contaminant level; USEPA, U.S. Environmental Protection Agency]

Constituent	Benchmark type	Benchmark value	Number of grid wells with GAMA data	Number of grid wells where supplemental CDPH data were used
Nutrient				
Ammonia, as nitrogen	HAL-US	[1]24.7 mg/L	29	0
Nitrate plus nitrite, as nitrogen	MCL-US	10 mg/L	29	47
Nitrite, as nitrogen	MCL-US	1 mg/L	29	44
Trace element				
Aluminum	MCL-CA	1,000 µg/L	29	42
Antimony	MCL-US	6 µg/L	29	41
Arsenic	MCL-US	10 µg/L	29	42
Barium	MCL-CA	1,000 µg/L	29	42
Beryllium	MCL-US	4 µg/L	29	42
Boron	NL-CA	1,000 µg/L	29	25
Cadmium	MCL-US	5 µg/L	29	42
Chromium	MCL-CA	50 µg/L	29	42
Copper	AL-US	1,300 µg/L	29	41
Iron	SMCL-CA	300 µg/L	29	41
Lead	AL-US	15 µg/L	29	42
Manganese	SMCL-CA	50 µg/L	29	41
Mercury	MCL-US	2 µg/L	29	42
Molybdenum	HAL-US	40 µg/L	29	5
Nickel	MCL-CA	100 µg/L	29	42
Selenium	MCL-US	50 µg/L	29	42
Silver	SMCL-CA	100 µg/L	29	41
Strontium	HAL-US	4,000 µg/L	29	1
Thallium	MCL-US	2 µg/L	29	42
Uranium	MCL-US	30 µg/L	29	15
Vanadium	NL-CA	50 µg/L	29	26
Zinc	SMCL-US	5,000 µg/L	29	41
Minor ion				
Fluoride	MCL-CA	2 mg/L	29	42
Major ion				
Chloride	SMCL-CA	500 mg/L	29	41
Sulfate	SMCL-CA	500 mg/L	29	41
Total dissolved solids (TDS)	SMCL-CA	1,000 mg/L	29	41
Radioactive				
Gross-alpha radioactivity, 72 hour count	MCL-US	15 pCi/L	29	38
Gross-beta radioactivity, 72 hour count	MCL-CA	50 pCi/L	29	3
Radium-226	MCL-US	5 pCi/L	29	0
Radium-228	MCL-US	5 pCi/L	29	18
Radon-222	MCL-US	4,000 pCi/L	29	1

[1]The HAL-US is 30 mg/L "as ammonia." To facilitate comparison to the analytical results, we have converted and reported this HAL-US as 24.7 mg/L "as nitrogen."

Table 3. Comparison of the number of compounds and median laboratory reporting levels or method detection limits by type of constituent for data reported in the California Department of Public Health (CDPH) database and for data collected by the U.S. Geological Survey (USGS) for the Monterey Bay and Salinas Valley Basins study unit, California GAMA Priority Basin Project, July–October 2005.

[µg/L, Micrograms per liter; mg/L, milligrams per liter; pCi/L, picocuries per liter; MDL, method detection limit; LRL, laboratory reporting level; ssL_C, sample-specific critical level; SSMDC, sample-specific minimum detectable concentration; ns, not sampled]

Constituent	CDPH		USGS GAMA		Median unit
	Number of compounds	Median MDL	Number of compounds	Median LRL	
Organic constituents					
Volatile organic compounds (VOCs) plus gasoline additives (including fumigants)	61	0.5	88	0.06	µg/L
Pesticides plus degradates	27	2	114	0.019	µg/L
Inorganic constituents					
Nutrients, major and minor ions	4	0.4	17	0.06	mg/L
Trace elements	20	8	25	0.12	µg/L
Radioactive constituents (ssL_C)[1]	5	1	8	[1] 0.04	pCi/L
Constituents of special interest					
Perchlorate	1	4	1	0.5	µg/L
N-Nitrosodimethylamine (NDMA)	ns	ns	1	0.002	µg/L

[1]The median laboratory reporting level used by USGS-GAMA for radioactive constituents is the sample-specific critical level, ssL_C.

Selection of Constituents for Additional Evaluation

As many as 284 constituents were analyzed in samples from MS study unit wells; however, only a subset of these constituents is discussed in this report. Three criteria were used to select constituents for additional evaluation:

1. Constituents present at high or moderate relative-concentrations in the CDPH database within the 3-year interval (July 17, 2002–July 18, 2005);

2. Constituents present at high or moderate relative-concentrations in the USGS-grid wells or USGS-understanding wells; or

3. Organic constituents with detection frequencies of greater than 10 percent in the USGS-grid well dataset for the study unit.

These criteria identified 7 organic constituents and 16 inorganic constituents for additional evaluation in the *status assessment*. An additional 25 organic constituents and 34 inorganic constituents were detected by USGS-GAMA, but were not selected for additional evaluation in the *status assessment* because either benchmarks were not established or detection was at low relative-concentrations (table 4).

Constituents discussed in the *understanding assessment* had high relative-concentrations in greater than 2 percent of the primary aquifers, or were detected in greater than 10 percent of the USGS-grid well dataset. A complete list of the constituents investigated by USGS-GAMA in the Monterey Bay and Salinas Valley Basins study unit may be found in the MS Data Report (Kulongoski and Belitz, 2007).

The CDPH database also was used to identify constituents with high relative-concentrations historically, but not currently. The historical period was defined as from the earliest record maintained in the CDPH database to July 17, 2002 (April 24, 1974–July 17, 2002).

Constituent concentrations may be historically high, but not currently high, because of improvement of groundwater quality with time or abandonment of wells with high concentrations. Historically high concentrations of constituents that do not otherwise meet the criteria for additional evaluation in the *status assessment* are not considered representative of potential groundwater-quality concerns in the study unit from 2002 to 2005. For the MS study unit, 20 constituents were measured at high relative-concentrations prior to July 17, 2002 (table 5).

Table 4. Aquifer-scale proportions from grid-based and spatially weighted approaches for constituents with high relative-concentrations during July 17, 2002–July 18, 2005 from the California Department of Public Health (CDPH) database, or with moderate or high relative-concentrations in samples collected from USGS-grid wells (July–October 2005), Monterey Bay and Salinas Valley Basins study unit, California GAMA Priority Basin Project.

[Aquifer-scale proportions are areal. Benchmark value units: trace elements, solvents, gas additives, trihalomethanes, herbicides, fumigants, insecticides, and NDMA, micrograms per liter (µg/L); radioactive constituents, picocuries per liter (pCi/L); nutrients, milligrams per liter (mg/L). Grid-based aquifer-scale proportions for organic constituents are based on samples collected by the U.S. Geological Survey from 91 grid wells during July to October 2005. Spatially weighted aquifer proportions are based on CDPH data from July 17, 2002, to July 18, 2005, in combination with grid well and understanding well data. High, concentrations greater than benchmark; moderate, concentrations less than benchmark and greater than or equal to 0.1 (for organic constituents) or 0.5 (for inorganic constituents) of benchmark; low, concentrations less than 0.1 (for organic constituents) or 0.5 (for inorganic constituents) of benchmark. MCL-CA, CDPH maximum contaminant level; MCL-US, USEPA maximum contaminant level; NL-CA, CDPH notification level; AL-US, USEPA action level; HAL-US, USEPA lifetime health advisory level; SMCL-CA, CDPH secondary maximum contaminant level; SMCL-US, USEPA secondary maximum contaminant level; RSD5, USEPA risk specific dose at 10⁻⁵; CDPH, California Department of Public Health; USEPA, U.S. Environmental Protection Agency; na, data not available; ns, not sampled]

Constituent	Benchmark		Raw detection frequency[1]			Spatially weighted aquifer-scale proportion of moderate and high relative-concentrations[1]			Grid-based aquifer-scale proportion of moderate and high relative-concentrations			90-percent confidence interval for grid-based high proportion[2]	
	Type	Value	Number of wells	Moderate values (percent)	High values (percent)	Number of cells	Moderate values (percent)	High values (percent)	Number of cells	Moderate values (percent)	High values (percent)	Lower limit (percent)	Upper limit (percent)
Trace elements													
Aluminum	MCL-CA	1,000	297	0.7	0.3	72	1.4	0.1	71	1.4	0.0	0.0	2.3
Antimony[3]	MCL-US	6	284	0.0	0.0	71	0.0	0.0	70	0.0	0.0	0.0	2.3
Arsenic	MCL-US	10	307	8.8	5.9	72	8.6	4.4	71	9.9	2.8	0.9	8.1
Barium	MCL-CA	1,000	298	0.3	0.0	72	0.5	0.0	71	1.4	0.0	0.0	2.3
Boron	NL-CA	1,000	216	6.5	0.5	62	4.3	1.6	54	7.4	1.9	0.4	8.0
Cadmium	MCL-US	5	299	0.7	0.3	72	1.7	0.3	71	2.8	0.0	0.0	2.3
Chromium	MCL-CA	50	277	0.7	0.0	72	0.4	0.0	71	0.0	0.0	0.0	2.3
Copper	AL-US	1,300	289	0.3	0.0	70	0.4	0.0	70	1.4	0.0	0.0	2.3
Lead	AL-US	15	297	1.0	0.7	72	0.4	0.9	71	0.0	0.0	0.0	2.3
Molybdenum	HAL-US	40	49	6.1	6.1	33	5.2	4.0	34	5.9	2.9	0.6	12.1
Selenium[3]	MCL-US	50	298	0.7	0.0	72	0.4	0.0	71	0.0	0.0	0.0	2.3
Uranium[3]	MCL-US	30	92	6.5	0.0	47	6.4	0.0	44	9.1	0.0	0.0	3.6
Vanadium	NL-CA	50	201	3.5	0.0	60	1.8	0.0	55	1.8	0.0	0.0	2.9
Trace element with SMCL													
Iron	SMCL-CA	300	292	7.2	13.0	70	7.1	12.1	70	7.1	7.1	3.5	14.0
Manganese	SMCL-US	50	292	4.5	16.8	70	5.5	17.9	70	4.3	18.6	12.1	27.3
Zinc	SMCL-US	5,000	288	0.3	0.0	70	0.3	0.0	70	0.0	0.0	0.0	2.3
Radioactive constituent													
Gross-alpha radioactivity, 72-hour count	MCL-US	15	301	5.3	1.0	68	7.3	1.7	67	10.4	1.5	0.3	6.4
Combined radium-226 and radium-228[3]	MCL-US	5	121	0.0	0.0	52	0.0	0.0	29	0.0	0.0	0.0	5.4
Nutrient													
Nitrate, as nitrogen	MCL-US	10	393	10.2	5.9	77	5.6	6.3	76	5.3	7.9	4.1	14.5
Nitrite, as nitrogen	MCL-US	1	305	0.3	0.0	74	0.5	0.0	73	1.4	0.0	0.0	2.2

Table 4. Aquifer-scale proportions from grid-based and spatially weighted approaches for constituents with high relative-concentrations during July 17, 2002–July 18, 2005 from the California Department of Public Health (CDPH) database, or with moderate or high relative-concentrations in samples collected from USGS-grid wells (July–October 2005), Monterey Bay and Salinas Valley Basins study unit, California GAMA Priority Basin Project.—Continued

[Aquifer-scale proportions are areal. Benchmark value units: trace elements, solvents, gas additives, trihalomethanes, herbicides, fumigants, insecticides, and NDMA, micrograms per liter (μg/L); radioactive constituents, picocuries per liter (pCi/L); nutrients, milligrams per liter (mg/L). Grid-based aquifer-scale proportions for organic constituents are based on samples collected by the U.S. Geological Survey from 91 grid wells during July to October 2005. Spatially weighted aquifer proportions are based on CDPH data from July 17, 2002, to July 18, 2005, in combination with grid well and understanding well data. High, concentrations greater than benchmark; moderate, concentrations less than benchmark and greater than or equal to 0.1 (for organic constituents) or 0.5 (for inorganic constituents) of benchmark; low, concentrations less than 0.1 (for organic constituents) or 0.5 (for inorganic constituents) of benchmark. MCL-CA, CDPH maximum contaminant level; MCL-US; USEPA maximum contaminant level; NL-CA, CDPH notification level; AL-US, USEPA action level; HAL-US, USEPA lifetime health advisory level; SMCL-CA, CDPH secondary maximum contaminant level; SMCL-US, USEPA secondary maximum contaminant level; RSD5, USEPA risk specific dose at 10^{-5}; CDPH, California Department of Public Health; USEPA, U.S. Environmental Protection Agency; na, data not available; ns, not sampled]

Constituent	Benchmark		Raw detection frequency[1]			Spatially weighted aquifer-scale proportion of moderate and high relative-concentrations[1]			Grid-based aquifer-scale proportion of moderate and high relative-concentrations			90-percent confidence interval for grid-based high proportion[2]	
	Type	Value	Number of wells	Moderate values (percent)	High values (percent)	Number of cells	Moderate values (percent)	High values (percent)	Number of cells	Moderate values (percent)	High values (percent)	Lower limit (percent)	Upper limit (percent)
Major and minor ion and total dissolved solids (TDS)													
Chloride	SMCL-CA	500	288	2.1	0.3	70	2.4	1.4	70	0.0	1.4	0.3	6.2
Sulfate	SMCL-CA	500	288	4.2	0.7	70	7.4	2.9	70	8.6	2.9	1.0	8.3
Total dissolved solids (TDS)	SMCL-CA	1,000	290	37.2	4.1	70	31.5	8.7	70	31.4	8.6	4.5	15.7
Solvent													
Carbon tetrachloride	MCL-CA	0.5	368	0.5	0.0	92	1.1	0.0	91	2.2	0.0	0.0	1.8
1,4-Dioxane[3]	NL-CA	3	4	na	0.0	2	na	0.0	0	ns	ns	ns	ns
Tetrachloroethene (PCE)	MCL-US	5	366	1.1	0.3	90	0.0	0.1	89	1.1	0.0	0.0	1.8
Trichloroethene (TCE)	MCL-US	5	368	0.8	0.0	92	0.6	0.0	91	1.1	0.0	0.0	1.8
Gasoline additive													
Methyl-*tert*-butyl-ether (MTBE)	MCL-CA	13	391	0.8	0.3	93	0.8	0.1	91	2.2	0.0	0.0	1.8
Trihalomethane													
Bromodichloromethane[3] (THM)	MCL-US	80	369	0.0	0.0	92	0.0	0.0	91	0.0	0.0	0.0	1.8
Chloroform[3]	MCL-US	80	355	0.3	0.0	88	0.4	0.0	76	0.0	0.0	0.0	2.1
Herbicide and fumigant													
Simazine	MCL-CA	4	274	0.0	0.0	92	0.0	0.0	91	0.0	0.0	0.0	1.8
1,4-Dichlorobenzene[3]	MCL-CA	5	368	0.0	0.0	92	0.0	0.0	91	0.0	0.0	0.0	1.8
Insecticide													
Dieldrin	RSD5-US	0.02	187	0.5	0.0	91	0.2	0.0	91	1.1	0.0	0.0	1.8
Special-interest constituent													
N-Nitrosodimethylamine (NDMA)	NL-CA	0.01	na	na	na	na	na	na	29	6.9	0.0	0.0	18.8

[1] Based on the most recent data for each CDPH well during July 17, 2002–July 18, 2005, combined with GAMA grid and understanding well data.

[2] Based on the Jeffrey's interval for the binomial distribution (Brown and others, 2001).

[3] The high value was reported in the CDPH database between July 17, 2002, and July 18, 2005, but this high value was not the most recently reported value used for calculating aquifer-scale proportion.

Table 5.   Constituents in the California Department of Public Health (CDPH) database at high concentrations from April 24, 1974–July 17, 2002, Monterey Bay and Salinas Valley Basins study unit, California GAMA Priority Basin Project.

[Benchmark value units in micrograms per liter (μg/L). MCL-CA, CDPH maximum contaminant level; NL-CA, CDPH notification level; MCL-US; USEPA maximum contaminant level; HAL-US, USEPA action level; CDPH, California Department of Public Health; USEPA, U.S. Environmental Protection Agency]

Constituent	Benchmark type	Benchmark value	Date of most recent high value	Number of wells with historically high values
Trace elements				
Chromium	MCL-CA	50	06-03-02	5
Fluoride	MCL-CA	2	04-16-91	4
Mercury	MCL-CA	2	01-06-86	3
Vanadium	NL-CA	50	06-25-02	2
Trace elements with SMCL				
Zinc	SMCL-CA	5,000	11-26-01	1
Solvents				
1,1-dichloroethane	MCL-CA	5	07-11-89	1
Dichloromethane (methylene chloride)	MCL-US	5	04-03-02	2
1,2-dichloroethane	MCL-CA	0.5	05-07-96	3
1,1,2,2-tetrachloroethane	MCL-CA	1	03-09-88	1
1,2,4-trichlorobenzene	MCL-CA	5	08-13-91	1
Other organic compounds				
Benzene	MCL-CA	1	10-05-98	4
Bromomethane	HAL-US	10	08-27-01	1
1,1-dichloroethene	MCL-CA	6	12-04-97	2
Di(2-ethylhexyl)phthalate	MCL-CA	4	10-21-98	2
Ethylene dibromide (EDB)	MCL-US	0.05	04-30-02	1
Naphthalene	NL-CA	17	09-07-93	1
Toluene	MCL-CA	150	12-07-93	1
Vinyl chloride	MCL-CA	0.5	07-05-89	1
Herbicides				
Atrazine	MCL-CA	1	07-16-02	1
Insecticides				
Heptachlor	MCL-CA	0.01	03-30-00	1

Calculation of Aquifer-Scale Proportions

Aquifer-scale proportions are defined as the percentage of the area (rather than the volume) of the primary aquifer system with concentrations greater than or less than specified thresholds relative to regulatory or aesthetic water-quality benchmarks. Two statistical approaches were selected to evaluate the proportions of the primary aquifers (Belitz and others, 2010) in the MS study unit with high, moderate, or low relative-concentrations of constituents relative to benchmarks:

- Grid-based: One value per grid cell from either USGS-grid or CDPH-grid wells was used to represent the primary aquifer system. The proportion of the primary aquifer system with high relative-concentrations was calculated by dividing the number of grid cells represented by a high relative-concentration for a particular constituent by the total number of grid cells with data for that constituent (see appendix C for details of methods). Proportions of moderate and low relative-concentrations were calculated similarly. Confidence intervals for grid-based detection frequencies of high concentrations were computed by using the Jeffreys interval for the binomial distribution (Brown and others, 2001). The grid-based estimate is spatially unbiased. However, the grid-based approach may not identify constituents that are present at high relative-concentrations in small proportions of the primary aquifers.

- Spatially weighted: All available data from the following sources were used to calculate the aquifer-scale proportions—all CDPH wells in the study unit (most recent analysis that passes the quality control tests from each well with data for that constituent during the current period (July 17, 2002, to July 18, 2005), USGS-grid wells, and USGS-understanding wells with perforation depth intervals representative of the primary aquifer system. For the spatially weighted approach, proportions were computed on a cell-by-cell basis (Isaaks and Srivastava, 1989) rather than as an average of all wells. The proportion of high relative-concentrations for each constituent for the primary aquifers was computed by (1) calculating the proportion of wells with high relative-concentrations in each grid-cell; and (2) averaging together the grid-cell proportions computed in step (1) (see appendix C for details of methods). Similar procedures were used to calculate the proportions of moderate and low relative-concentrations of constituents. The resulting proportions are spatially unbiased (Isaaks and Srivastava, 1989).

In addition, for each constituent, the detection frequencies of high and moderate relative-concentrations for individual constituents were calculated using the same dataset as used for the spatially weighted calculations. However, these "raw" detection frequencies are not spatially unbiased because the wells in the CDPH database are not uniformly distributed throughout the MS study unit (fig. 4). Consequently, high relative-concentrations in wells clustered in a particular area representing a small part of the primary aquifers could be given a disproportionately high weight compared to spatially unbiased methods. Raw detection frequencies are provided for reference but were not used to characterize the groundwater resource (see appendix C for details of statistical methods).

Aquifer-scale proportions discussed in this report primarily were estimated using the grid-based approach, and secondarily using the spatially weighted approach. The grid-based aquifer-scale proportions were used unless the spatially weighted proportions were significantly different. Significantly different results were defined as follows:

1. If the aquifer proportion for the high category was zero using the grid-based approach and non-zero using the spatially weighted approach, then the result from the spatially weighted approach was used. This situation can arise when the concentration of a constituent is high in a small fraction of the primary aquifers.

2. If the grid-based aquifer proportion for the high category was non-zero, then the 90 percent confidence interval (based on the Jeffreys interval for the binomial distribution, Brown and others, 2001) was used to evaluate the difference. If the spatially weighted proportion was within the 90 percent confidence interval, then the grid-based proportion was used. If the spatially weighted proportion was outside the 90 percent confidence interval, then the spatially weighted proportion was used.

Aquifer-scale proportions for the moderate and low categories primarily were determined from the grid-based estimates because for some constituents the reporting levels for analyses in CDPH were too high to distinguish between moderate and low relative-concentrations using the spatially-weighted approach.

Aquifer-scale proportions of high relative-concentrations also were determined for classes of constituents. The classes of organic constituents for which aquifer-scale proportions were calculated include solvents, gasoline additives, trihalomethanes, other organic constituents, herbicides and fumigants, insecticides, and special-interest constituents. The classes of inorganic constituents with human-health benchmarks for which aquifer-scale proportions were calculated include trace elements, radioactive constituents, and nutrients. Classes of inorganic constituents with aesthetic benchmarks, for which aquifer-scale proportions were calculated, include major and minor ions (which include sulfate and chloride), total dissolved solids (TDS), trace elements, and manganese and iron.

Understanding-Assessment Methods

Potential explanatory factors—land use, well depth, depth to the top-of-perforation of the well, normalized position of wells along flowpaths, geochemical condition, and groundwater-age class (see appendix A for more details)—were analyzed in relation to constituents of interest for the *understanding assessment* in order to establish context for physical and chemical processes within the groundwater system. Statistical tests were used to identify significant correlations between the constituents of interest and potential explanatory factors. The strongest correlations for understanding factors influencing water quality are shown graphically.

The wells selected for the *understanding assessment* were USGS-grid and CDPH-grid wells, and USGS-understanding wells. CDPH "DPH" wells were not used in the *understanding assessment* because carbon isotope, tritium, dissolved oxygen, and some well construction data were not available. Correlations between water-quality variables and potential explanatory factors were tested using either the set of grid and understanding wells combined or grid wells only. Because the USGS-understanding wells were not randomly selected on a spatially distributed grid, these wells were excluded from analyses of relations of water quality to areally distributed factors (land use) to avoid areal-clustering bias. However, six USGS-understanding wells were selected for analyses of relations between constituents and the vertically distributed explanatory factors (depth, groundwater-age classification, and geochemical conditions). TDS was measured directly or calculated from specific conductance (see appendix D).

Statistical Analysis

Nonparametric statistical methods were used to test the significance of correlations between water-quality variables and potential explanatory factors. Nonparametric statistics are robust techniques that generally are not affected by outliers and do not require that the data follow any particular distribution (Helsel and Hirsch, 2002). The significance level (p) used to test hypotheses for this report was compared to a threshold value (α) of 5 percent (α = 0.05) to evaluate whether the relation was statistically significant (p < α).

Correlations were investigated by using Spearman's method to calculate the rank-order correlation coefficient (ρ) between continuous variables. The values of ρ can range from +1.0 (perfect positive correlation) to 0.0 (no correlation) to -1.0 (perfect negative correlation). The Wilcoxon rank-sum test was used to evaluate the correlation between water quality

and categorical explanatory factors [for example, groundwater age (modern, mixed, or pre-modern), redox conditions (oxic, mixed, or anoxic/suboxic), and land-use classification (natural, agricultural, urban, or mixed)]. The Wilcoxon rank-sum test can be used to compare two independent populations (data groups or categories) to determine whether one population contains larger values than the other (Helsel and Hirsch, 2002). Correlations were investigated using the Wilcoxon rank-sum tests with exact distribution and continuity correction. A positive value of Z (Wilcoxon test statistic) indicates that the first classification is larger than the second, whereas a negative Z value indicates the first classification is smaller than the second. The null hypothesis for the Wilcoxon rank-sum test is that there is no significant difference between the values of the two independent data groups being tested.

The Wilcoxon rank-sum test was used for multiple comparisons of two independent groups rather than the multiple-stage Kruskal-Wallis test for identifying differences between three or more groups, although a set of Wilcoxon rank-sum tests is more likely to falsely indicate a significant difference between groups than the Kruskal-Wallis test (Helsel and Hirsch, 2002). However, given the potentially large and variable number of differences to be evaluated, the Wilcoxon rank-sum test was selected as a consistent and practical direct test of differences. Because of the small sample size, the exact distribution with continuity correction also was applied.

Potential Explanatory Factors

Land Use

Land use classified as natural made up the greatest percentage (43.7 percent) of the total land area in the study unit in 1992 (taking into account the entire area of the study unit, rather than just the area around grid wells), whereas agricultural land use in 1992 was 43.4 percent of the study unit area, and urban land use in 1992 was only 12.9 percent of the study unit area (figs. 5A and 6). However, land use in 1992 surrounding USGS-grid wells (500-meter radius (1,640 ft)) in the MS study unit was nearly equally distributed among land-use classifications (Nakagaki and others, 2007)—35.4 percent agricultural, 33.2 percent urban, and 31.4 percent natural (fig. 5A). In the MS study unit, natural lands are mostly grassland and forests, whereas the primary use of agricultural land is for row crops, pasture (cattle, sheep, and poultry), hay, and vineyards (Nakagaki and others, 2007). The largest urban areas are the cities of Santa Cruz, Watsonville, Monterey, Salinas, King City, and Paso Robles.

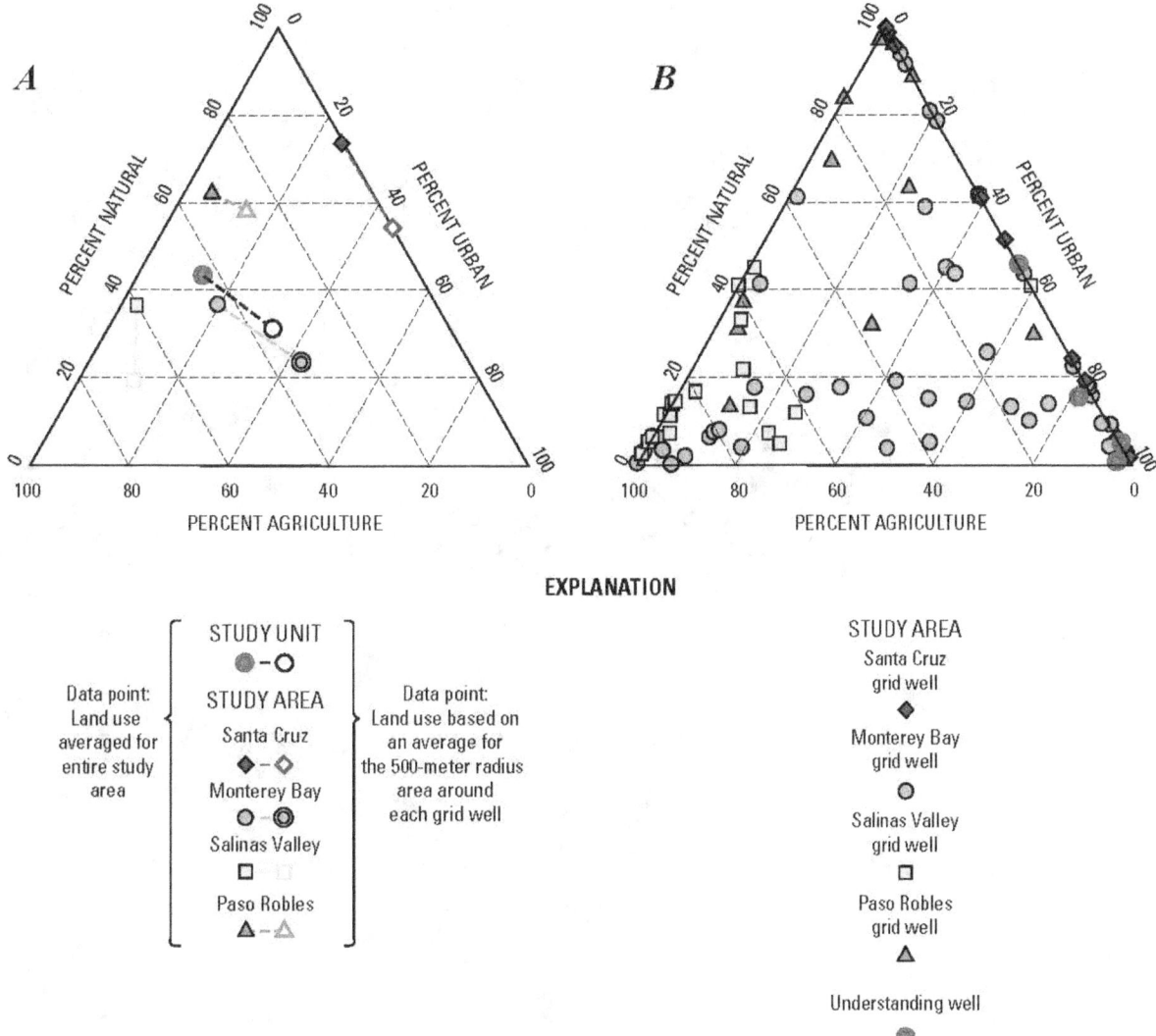

EXPLANATION

Figure 5.   Percentage of urban, agricultural, and natural land use in (*A*) the study unit and study areas, and (*B*) the area surrounding each grid and understanding well in the Monterey Bay and Salinas Valley Basins study unit, California GAMA Priority Basin Project.

Land use in the SV and MB study areas is predominantly agricultural, whereas the SC study area is predominantly natural (fig. 6). In the SC study area, 45 percent of land use within 500-m radius (1,640 ft) area surrounding each grid well was urban, but only 23 percent of the entire study area was urban (fig. 5*A*). The high percentage of urbanized land surrounding the grid wells—compared to the land in the entire study area—indicates the association of public-supply wells with population density. The area surrounding grid wells, particularly for the SC study area, may be influenced more heavily by urban activities than might be expected based on the average land use of the entire study area. A 500-m buffer surrounding the well has been shown to be effective at correlating urban land use with VOC occurrence, for the purposes of statistical characterization (Johnson and Belitz, 2009).

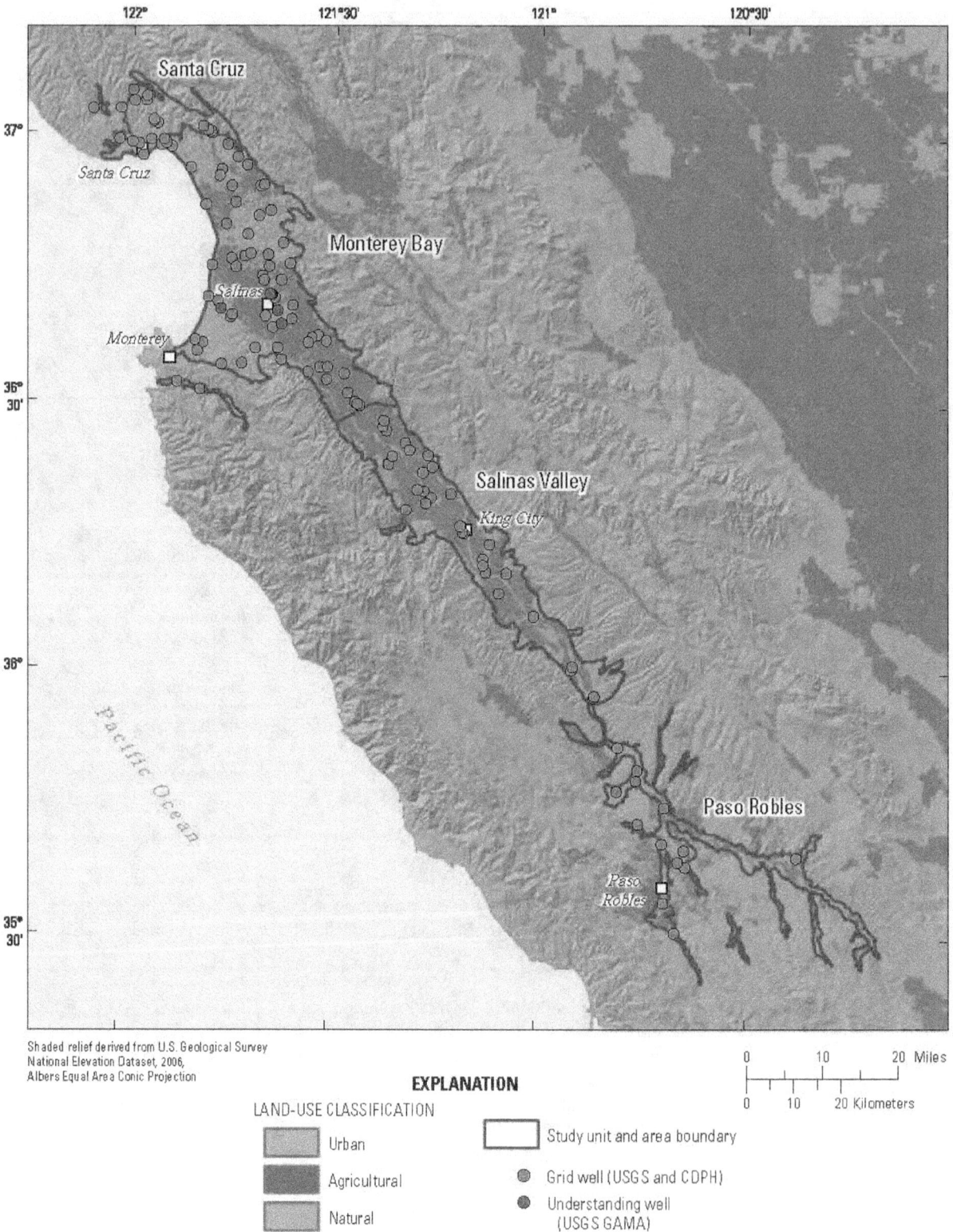

Shaded relief derived from U.S. Geological Survey
National Elevation Dataset, 2006,
Albers Equal Area Conic Projection

EXPLANATION

LAND-USE CLASSIFICATION

Urban

Agricultural

Natural

Study unit and area boundary

Grid well (USGS and CDPH)

Understanding well
(USGS GAMA)

Figure 6.   Land use in the Monterey Bay and Salinas Valley Basins study unit, California GAMA Priority Basin Project.

Well Depth and Depth to Top-of-Perforation

Well construction information was available for 81 of the 91 grid wells sampled in the MS study unit. Depths of grid wells ranged from 69 to 1,950 ft (21 to 595 m) below land surface (BLS); the median was 490 ft BLS (150 m) (fig. 7; table A1). Depths to the top-of-perforation ranged from 46 to 1,390 ft BLS (14 to 424 m), with a median of 232 ft BLS (71 m). The perforation length was as much as 970 ft (296 m) with a median of 200 ft (61 m). The wide range in construction depths reflects the geological differences between the SC, MB, SV, and PR study areas. Well depths and depth to top-of-perforations of understanding wells (three of six wells were public supply wells) were similar to those of the grid wells.

Normalized Position of Wells along Flowpath

Wells were sampled along the Salinas River Valley in order to assess how the positions of wells along flowpaths affected groundwater quality (see section, "Normalized Position of Wells along a Flowpath" in appendix A; table A1). This study examined the changes in concentrations of major and minor ions, and trace elements, as a function of the normalized position of wells along a groundwater flowpath. Types of wells considered for these flowpaths included grid, understanding, and CDPH "DPH" wells. There were 34 wells along the Salinas River Valley flowpath (figs. 8A, 8B).

Groundwater Age

Groundwater samples were assigned age classifications on the basis of the tritium, carbon-14, and helium-4 content of the samples (see section, "Groundwater Age Classification" in appendix A). Age classifications were assigned to 97 USGS-grid and understanding well samples; 27 were classified as modern, 27 were mixed (evidence of modern and pre-modern groundwater in the same sample), and 43 were pre-modern age (table A3).

Groundwater ages generally increased with depth to top of well perforations (fig. 9A). The depths to the top of perforations were significantly shallower in wells having water classified as modern age, compared to those classified as pre-modern age. The wells classified as modern were significantly shallower than wells classified as mixed or pre-modern (fig. 9B). Water in 6 of the 10 wells perforated entirely within the upper 200 ft (61 m) of the aquifer was modern age, whereas water in most wells (33 of 45) with perforations equal to or greater than 200 ft (61 m) below land surface was pre-modern (fig. 9C).

Geochemical Condition

An abridged classification of oxidation-reduction (redox) conditions adapted from the framework presented by McMahon and Chapelle (2008) was applied to data from 97 wells sampled by the USGS-GAMA Priority Basin Project, and to data from 16 wells reported in the CDPH database (appendix A; table A2). The classification "indeterminate" was added to the framework for groundwater samples that did not have sufficient data available to be classified as oxic, anoxic/suboxic, or mixed anoxic/oxic (Jurgens and others, 2009). Groundwater was oxic in 58 percent of the wells, mixed anoxic/oxic in 11 percent of the wells, and anoxic/suboxic (anoxic to suboxic) in 20 percent of the wells (table A2).

Correlations Between Explanatory Factors

Apparent correlations between explanatory factors and a water-quality constituent could actually reflect correlations between two or more explanatory factors. Therefore, it is important to identify statistically significant correlations between explanatory factors (table 6).

Land use in the MS study unit was not significantly correlated with any of the other explanatory variables. Depth to top-of-perforations had a significant positive correlation with normalized position along the flowpath and with well depth. This may reflect deep wells, and thus deep perforation intervals in wells, towards the distal end of the valleys. The depth of the wells had a significant positive correlation with normalized position of wells along the flowpath.

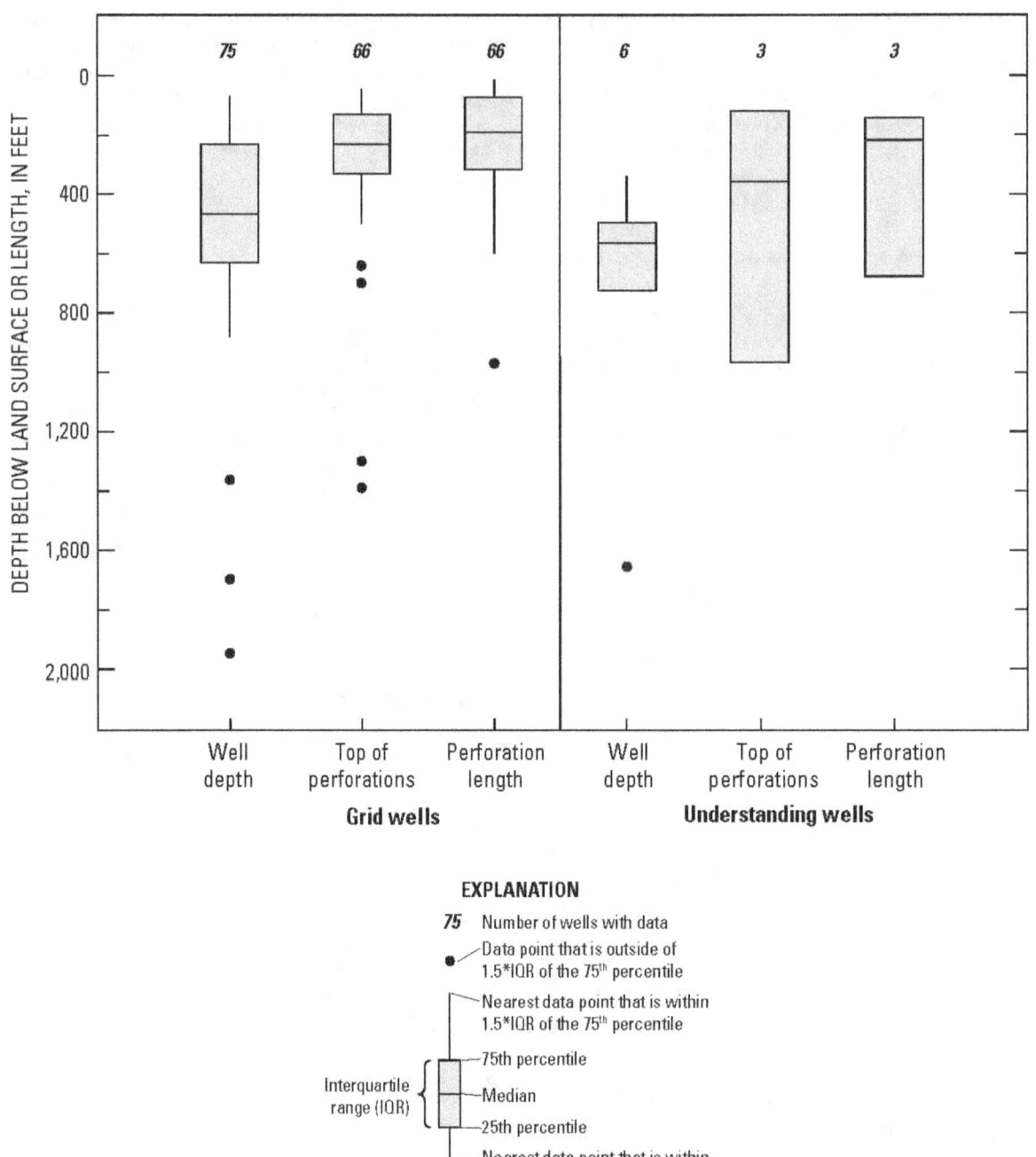

Figure 7.   Well depths, depths to top-of-perforation, and perforation lengths for grid and understanding wells, Monterey Bay and Salinas Valley Basins study unit, California GAMA Priority Basin Project.

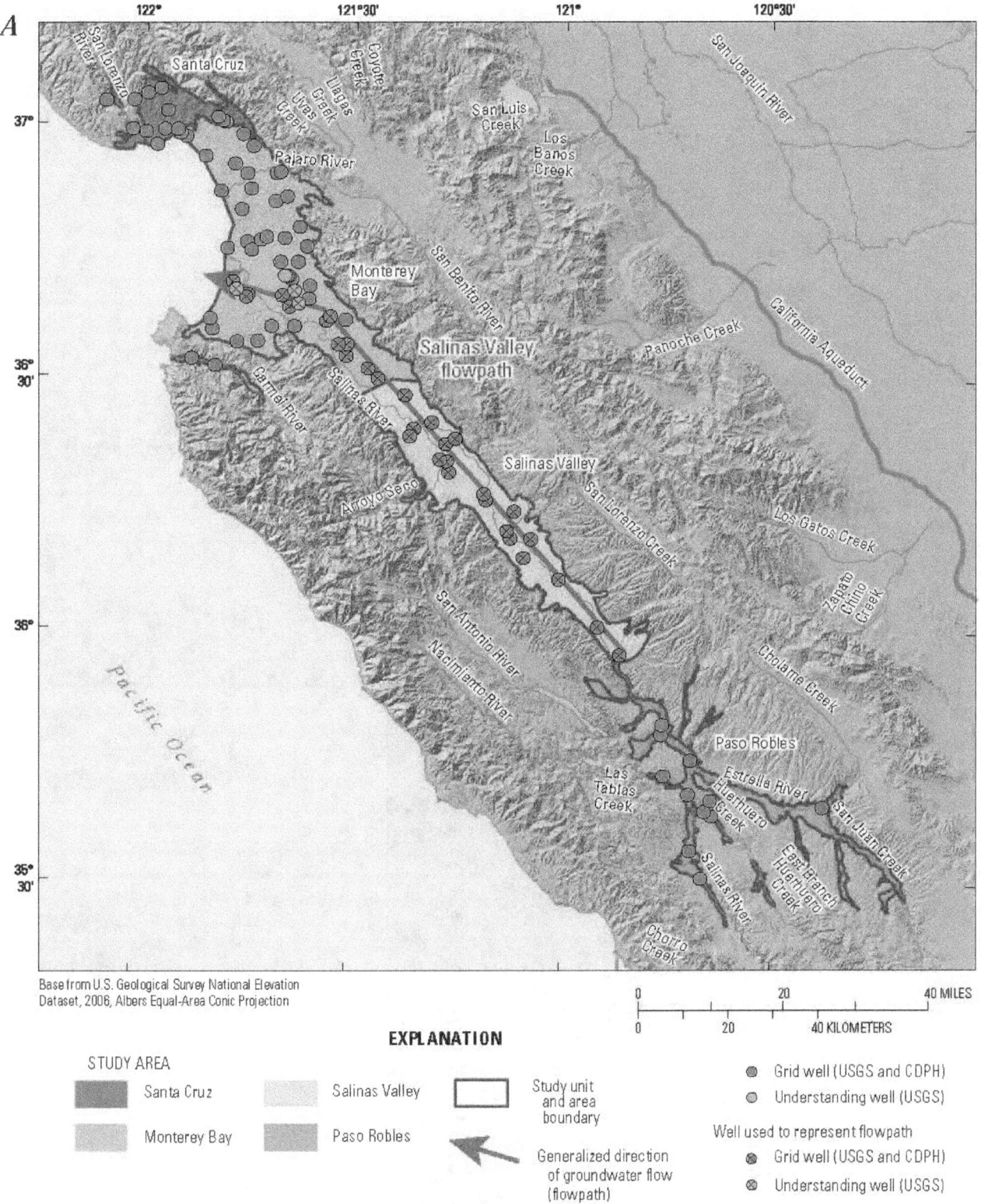

Figure 8. (A) Normalized position of wells along the Salinas Valley flowpath, and (B) conceptual model of the aquifer system in the Salinas Valley for the Monterey Bay and Salinas Valley Basins study unit, California GAMA Priority Basin Project.

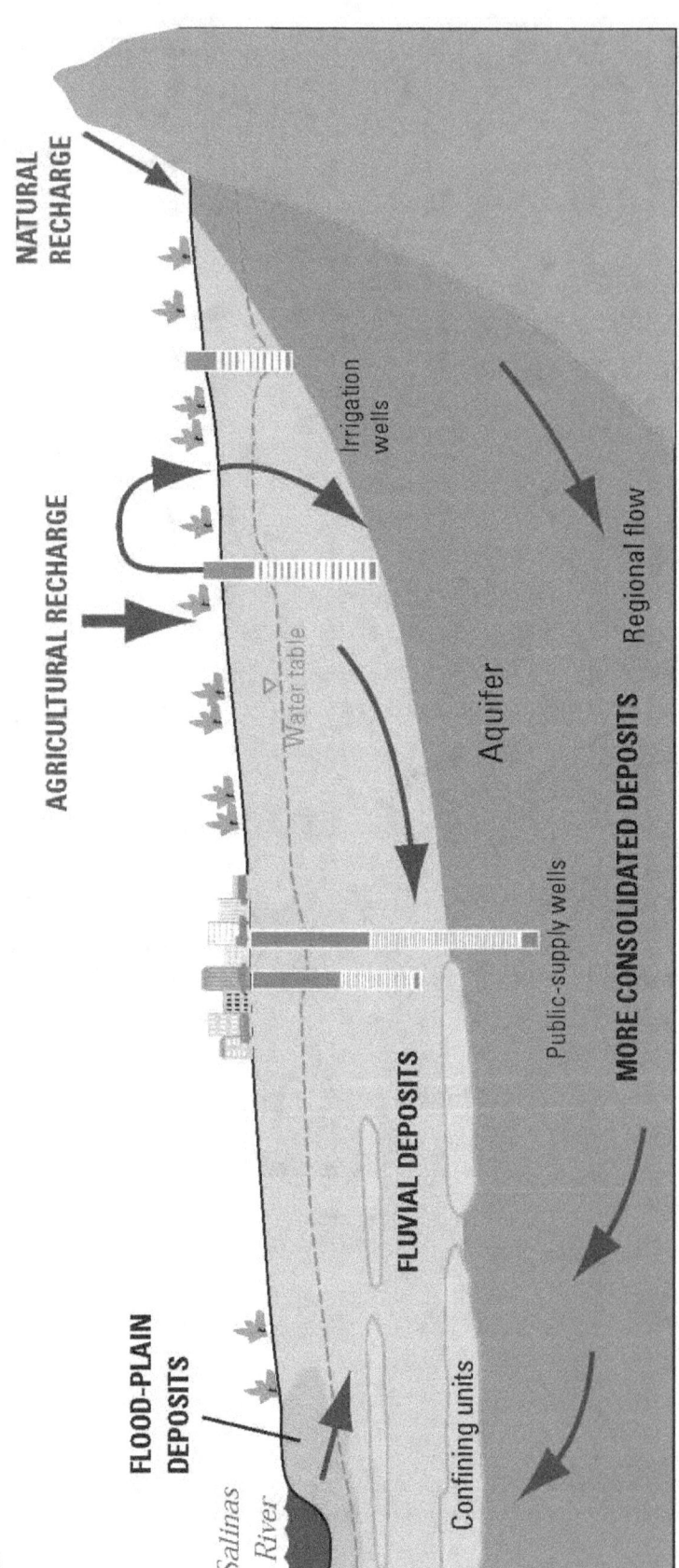

B

Figure 8.—Continued

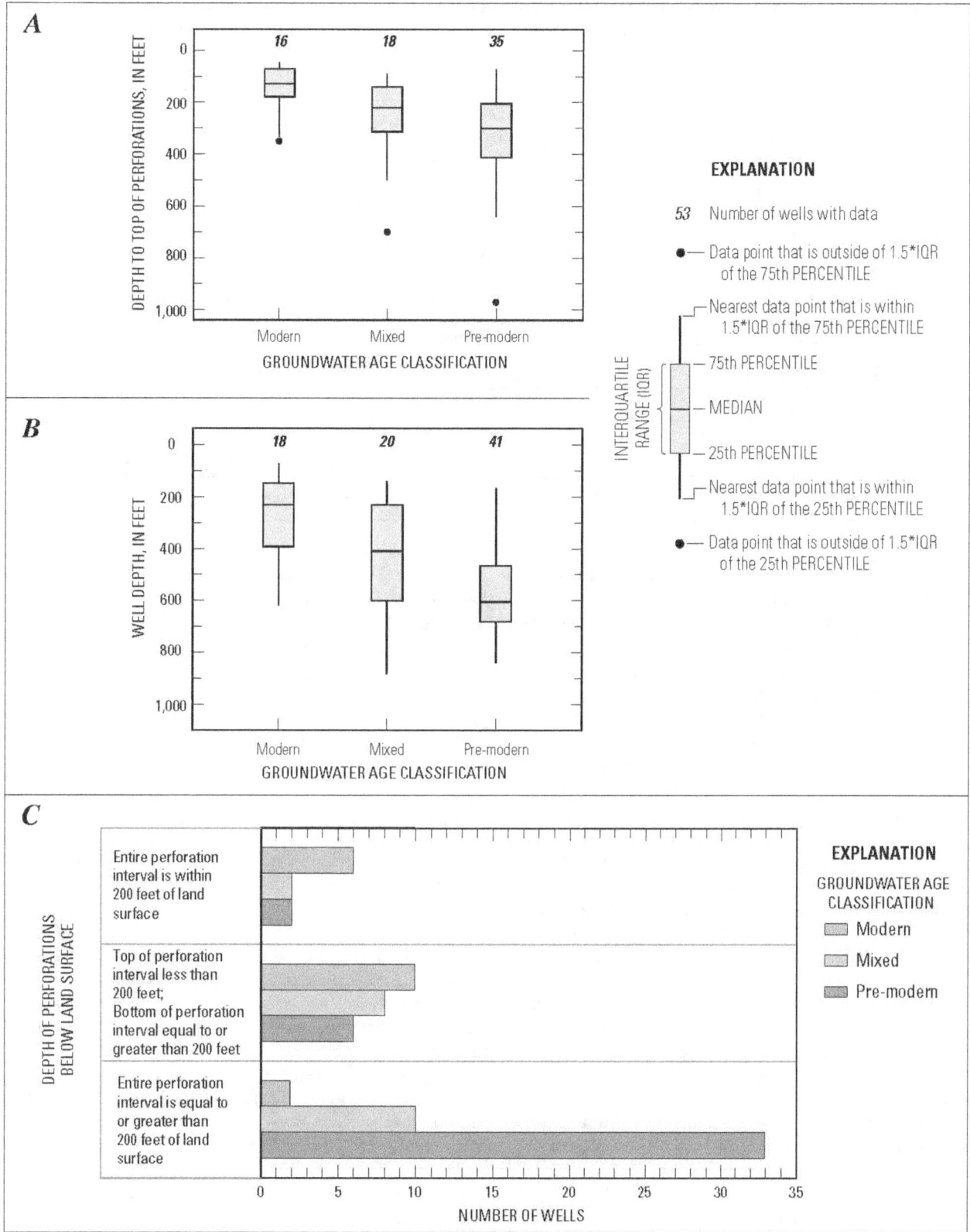

Figure 9. Relation of groundwater age classification to depth to top-of-perforations, well depth, and age classification, in relation to the depth of well perforations, Monterey Bay and Salinas Valley Basins study unit, California GAMA Priority Basin Project.

Table 6.   Results of non-parametric (Spearman's *rho* method) analysis of correlations in grid and understanding wells between selected potential explanatory factors, Monterey Bay and Salinas Valley Basins study unit, California GAMA Priority Basin Project.

[ρ, Spearman's correlation statistic; significant positive correlation and significant negative correlations shown; nc, no significant correlation]

Type of well analyzed	Explanatory factor	Normalized position of well along flowpath	ρ :Spearman's correlation statistic			
			Depth to top-of-perforations	Depth of well	Dissolved oxygen concentration	pH
Grid wells	Percentage of urban land use	nc	nc	nc	nc	nc
	Percentage of agricultural land use	nc	nc	nc	nc	nc
	Percentage of natural land use	nc	nc	nc	nc	nc
	Normalized position along flowpath		0.75	0.50	nc	nc
Grid and understanding wells	Depth to top-of-perforations	0.68		0.78	nc	nc
	Depth of well	0.60	0.78		nc	nc
	Dissolved oxygen concentration	nc	nc	nc		nc
	pH	nc	nc	nc	nc	

Figure 10 presents a cross section of well perforation intervals and redox classification plotted as normalized position of wells along the flowpath on the x-axis and as depth of the perforation interval on the y-axis (see appendix A for details). In many groundwater-flow systems, the relatively shallow, upgradient wells are typically oxic, trending towards more anoxic groundwater farther along the generalized flowpath (Kulongoski and others, 2010). This is not the case in the upper Salinas Valley, where the shallow groundwater in the upgradient (proximal) section of the flowpath is anoxic/suboxic and trends towards more oxic conditions farther along in the medial portion of the flowpath. The reducing conditions in the upgradient flow system may be explained by a subsurface structure, the Gabilan High, restricting flow about (about 1 mi southeast of King City), resulting in diminished groundwater flow (Durham, 1974). The transition to oxic conditions at King City may be explained by the confluence of the Salinas River and San Lorenzo Creek, where oxic water infiltrates, and (or) by the infiltration of water from uses related to urban irrigation or municipal discharge. The deep wells in the distal (downgradient) section of the flowpath have anoxic/suboxic conditions, which are expected of water that is older and deeper in the groundwater-flow system.

The pH ranged from 6.2 to 8.8 in the USGS-grid wells, USGS-understanding wells, and CDPH-other wells (fig. 11A). The relation between pH and well depth is shown in figure 11B. Two trends are apparent: (1) The pH of groundwater classified as pre-modern is slightly higher (median = 7.5; n = 33) than modern groundwater (median = 7.3; n = 11); (2) Groundwater classified as pre-modern has higher pH than mixed and modern age water; the higher pH water (pH > 7.4) was classified as predominantly pre-modern age and most was from wells deeper than 200 ft (61 m) (fig. 11B), whereas the lower pH groundwater was characterized as shallow and modern or mixed age. In alkaline groundwater conditions (pH> 8), trace elements may have a positive correlation with pH because some trace elements are desorbed from, or inhibited from adsorbing to, particle surfaces under these conditions.

Wilcoxon rank-sum tests were used to determine significant differences between selected water-quality constituents and potential explanatory factors. Arsenic concentrations were significantly lower in wells classified as having water of modern and mixed ages than in wells classified as having water of pre-modern age (table 7). Boron concentrations were significantly lower in groundwater with oxic conditions than in groundwater with anoxic/suboxic conditions. Iron and manganese concentrations were significantly lower in groundwater with oxic conditions than in groundwater with mixed and anoxic/suboxic conditions. Organic solvents concentrations in groundwater were significantly lower in wells in agricultural land-use areas than in wells in urban land-use areas. Simazine concentrations in groundwater were significantly lower in wells in natural land-use areas than in wells in agricultural land-use areas. Implications of correlations between explanatory variables are discussed later in the report as part of the analysis of factors affecting individual constituents.

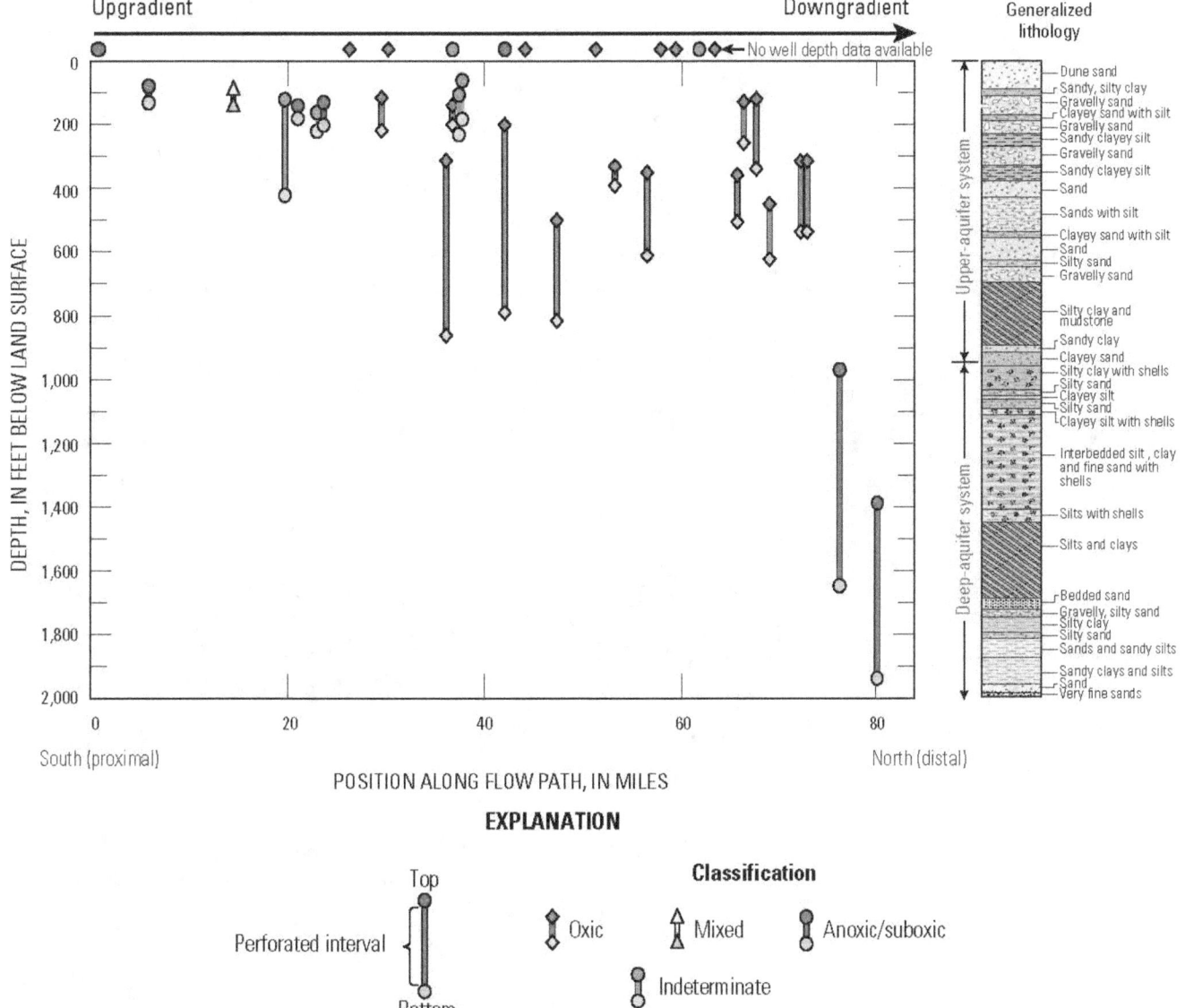

Figure 10. Relation of oxidation-reduction condition to normalized position of wells along a flowpath, and depth of perforated interval of wells, Monterey Bay and Salinas Valley Basins study unit, California GAMA Priority Basin Project.

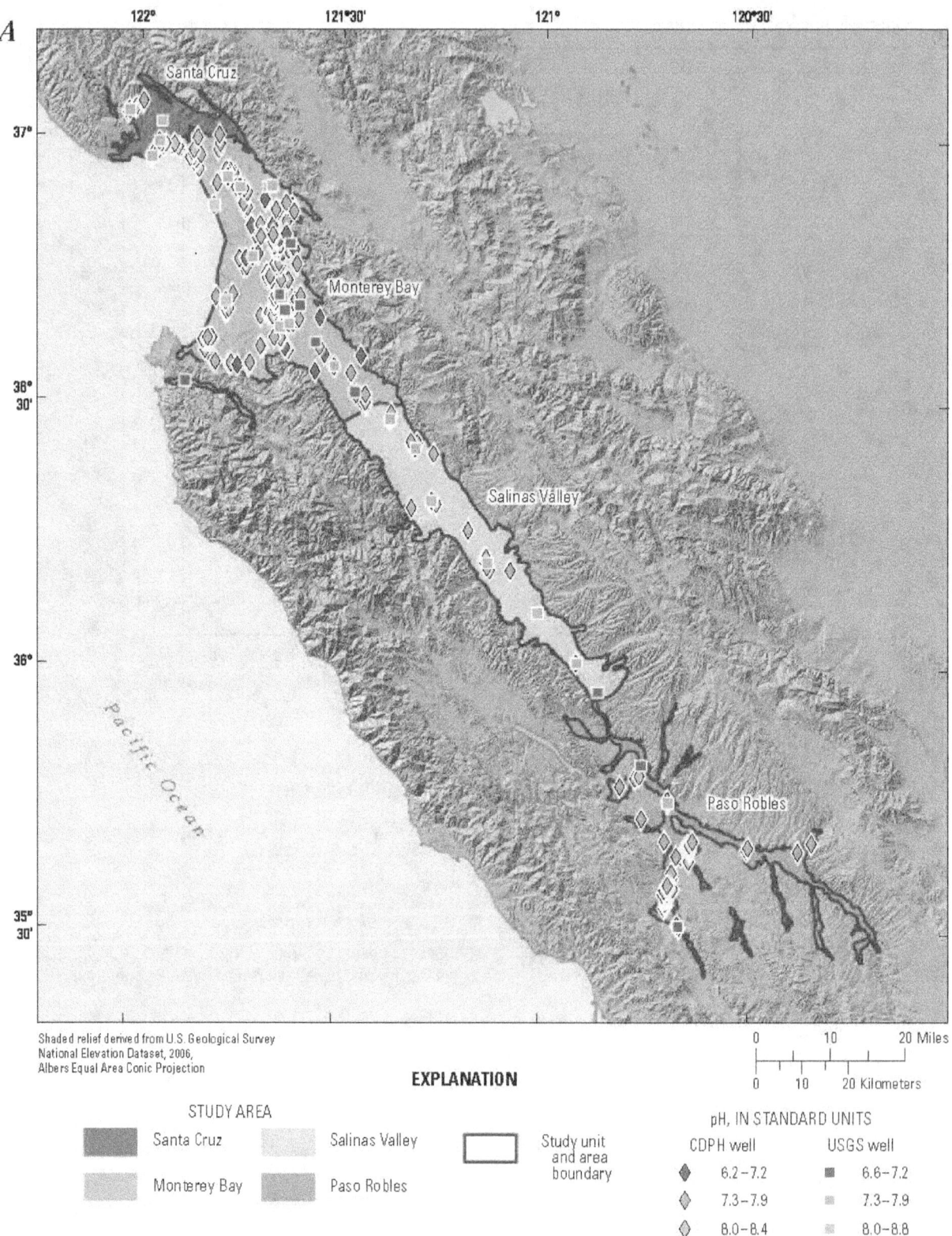

Figure 11. (A) pH levels in U.S. Geological Survey (USGS) wells and California Department of Public Health (CDPH) wells, and (B) graph showing pH plotted as a function of well depth and groundwater age classification, Monterey Bay and Salinas Valley Basins study unit, California GAMA Priority Basin Project.

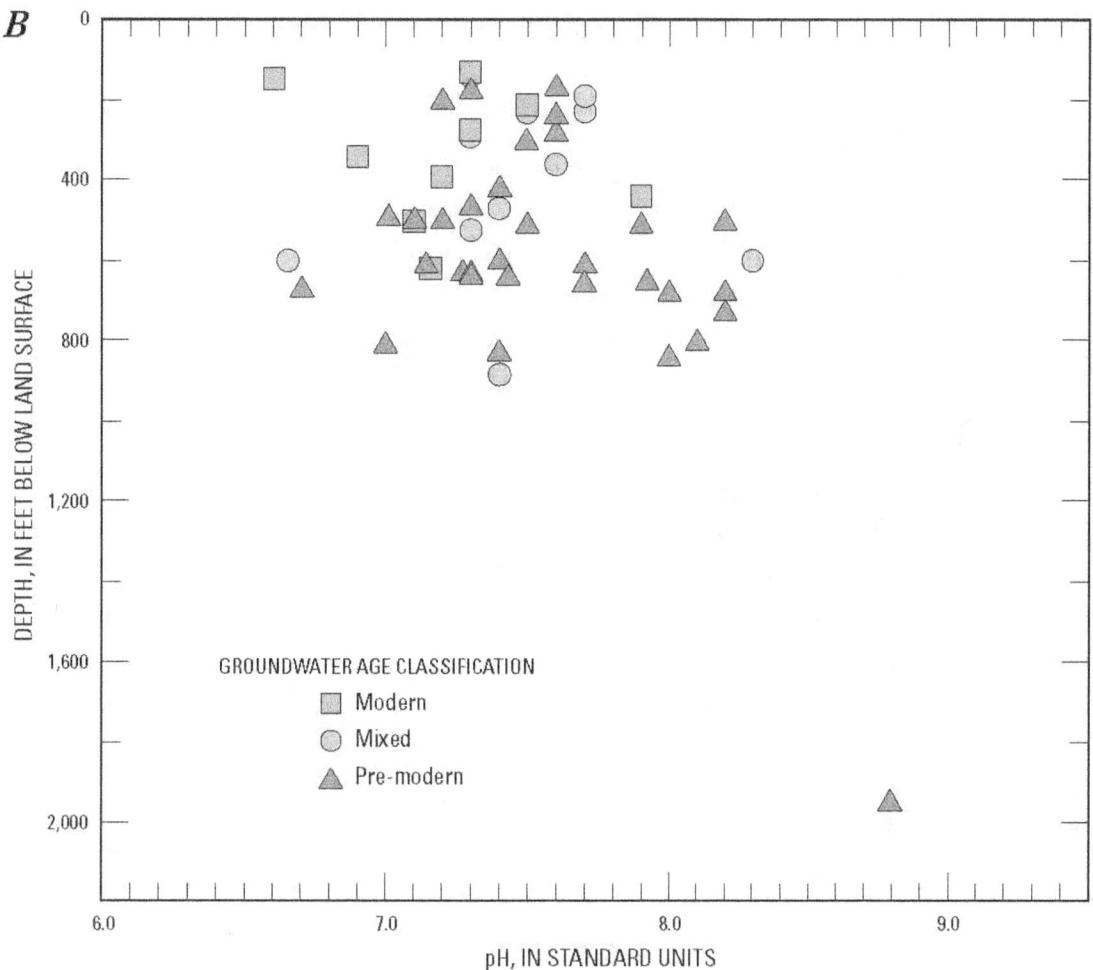

Figure 11.—Continued

Table 7. Results of Wilcoxon rank-sum tests on grid-well data used to determine significant differences between selected water-quality constituents grouped by potential explanatory factor classifications, Monterey Bay and Salinas Valley Basins study unit, California GAMA Priority Basin Project.

[Wilcoxon rank-sum tests with exact distribution and continuity correction; Z, test statistic for Wilcoxon test; significantly positive Z value (first classification is larger than second); significantly negative Z value (first classification is smaller than the second); PCE, tetrachloroethene; TCE, trichloroethene; nc, no significant correlation]

Selected water-quality constituent	Groundwater age classification			Geochemical conditions classification			Land-use classification					
	Modern compared with mixed	Mixed compared with pre-modern	Modern compared with pre-modern	Oxic compared with mixed	Mixed compared with anoxic/suboxic	Oxic compared with anoxic/suboxic	Natural compared with agricultural	Natural compared with mixed	Natural compared with urban	Agricultural compared with mixed	Agricultural compared with urban	Urban compared with mixed
						Z: Test statistic for Wilcoxon test						
Arsenic	nc	-2.50	-2.33	nc	nc	nc	nc	nc	nc	nc	nc	nc
Boron	nc	nc	nc	nc	nc	-2.34	nc	2.46	nc	2.34	nc	1.97
Gross alpha radioactivity	nc	nc	nc	nc	nc	nc	nc	nc	nc	nc	nc	nc
Molybdenum	nc	nc	nc	nc	nc	nc	nc	nc	1.96	2.22	nc	nc
Nitrate (as nitrogen)	nc	nc	nc	nc	nc	nc	-2.05	nc	nc	nc	nc	nc
Chloride	nc	nc	nc	nc	nc	nc	nc	2.03	nc	nc	nc	nc
Iron	nc	nc	nc	-4.67	nc	-4.85	nc	nc	nc	nc	nc	nc
Manganese	nc	nc	nc	-2.75	nc	-3.78	nc	nc	nc	nc	nc	nc
Sulfate	nc	nc	nc	nc	nc	nc	nc	2.23	nc	3.29	2.83	2.23
Total dissolved solids	nc	nc	nc	nc	nc	nc	nc	2.81	nc	2.86	nc	2.65
Dissolved oxygen	nc	nc	nc	3.23	4.46	6.88	nc	nc	nc	nc	nc	nc
pH	nc	nc	nc	nc	nc	nc	nc	nc	nc	nc	nc	nc
Solvents, sum of concentrations (PCE + TCE + carbon tetrachloride)	nc	nc	nc	nc	nc	nc	nc	nc	nc	nc	-3.68	nc
Simazine	nc	nc	nc	nc	nc	nc	-2.43	nc	nc	nc	nc	nc
N-Nitroso-dimethylamine (NDMA)	nc	nc	nc	nc	nc	nc	nc	nc	nc	nc	nc	nc

Status and Understanding of Water Quality

The *status assessment* was designed to identify the constituents or classes of constituents most likely to be of water-quality concern because of their high relative-concentrations or their prevalence. Approximately 23,000 individual analytical results were included in the assessment of groundwater quality for the MS study unit. The spatially distributed, randomized approach to grid-well selection and data analysis yields a view of groundwater quality in which all areas of the primary aquifers are weighted equally; regions with a high density of groundwater use or with high density of potential contaminants were not preferentially sampled (Belitz and others, 2010). The *understanding assessment* identifies the natural and human factors affecting water quality in the MS study unit, and focuses on the constituents selected for additional evaluation in the *status assessment*.

The following discussion of the *status* and *understanding assessment* results is divided into inorganic and organic constituents. The assessment begins with a survey of how many constituents were detected at any concentration compared to the number analyzed, and a graphical summary of the relative-concentrations of constituents detected in the grid wells. Results are presented for the subset of constituents that met criteria for selection for additional evaluation based on concentration, or for organic constituents, prevalence (see Selection of Constituents for Additional Evaluation).

The high aquifer-scale proportions calculated using the spatially weighted approach were within the 90-percent confidence intervals for their respective grid-based aquifer high proportions for 33 of the 34 constituents listed in table 4, providing evidence that the grid-based approach yields statistically equivalent results to the spatially weighted approach.

Inorganic Constituents

Inorganic constituents generally occur naturally in groundwater, although their concentrations may be influenced by human factors as well as natural factors. All 50 inorganic constituents analyzed by the USGS-GAMA were detected in the MS study unit, of which, 31 had regulatory or non-regulatory health-based benchmarks, 6 had non-regulatory aesthetic/technical-based benchmarks, and 13 had no established benchmarks (table 8). The inorganic constituents detected at high relative-concentrations in one or more of the 91 grid wells were arsenic, boron, molybdenum, iron, manganese, chloride, TDS, sulfate, nitrate, and gross alpha radioactivity (72-hour count). The maximum relative-concentration (sample concentration divided by the benchmark concentration) for each constituent is shown in figure 12.

Sixteen inorganic constituents—the trace elements aluminum, arsenic, barium, boron, cadmium, copper, molybdenum, uranium, vanadium, iron, and manganese; the major ions chloride and sulfate and TDS; gross alpha radioactivity (72-hour count); and the nutrient nitrate—met the selection criterion of having maximum relative-concentrations greater than 0.5 (moderate or high) in the grid-based aquifer-scale proportions (fig. 12) and are listed in table 4. Inorganic constituents, as a group (nutrients, trace elements, and radioactive constituents), had high relative-concentrations in 14.5 percent of the primary aquifers, moderate relative-concentrations in 35.5 percent, and low relative-concentrations in 50.0 percent (table 9).

Trace Elements

Trace elements, as a class, were detected at high relative-concentrations (for one or more constituents) in 5.6 percent of the primary aquifers, moderate values in 25.4 percent, and low values in 69.0 percent (table 9). High relative-concentrations of trace elements resulted from the high relative-concentrations of molybdenum (2.9 percent) and arsenic (2.8 percent) (table 4).

Inorganic constituents with relative-concentrations greater than 1.0 in one or more of the grid wells are shown in figure 13. The spatial distributions of selected inorganic constituents for USGS-grid wells and from the most recent years of available data (July 17, 2002–July 18, 2005) from the CDPH wells are shown in figures 14A–14H.

The percentage of the primary aquifer with high and moderate relative-concentrations for the individual constituents is shown in table 4. Molybdenum was detected at high relative-concentration in 2.9 percent of the primary aquifers and moderate relative-concentration in 5.9 percent (figs. 13 and 14A). Arsenic was detected at high relative-concentration in 2.8 percent of the primary aquifers and moderate relative-concentration in 9.9 percent. Relative-concentrations of arsenic were high in the MB and PR study areas, and moderate in the MB, SC, and PR study areas (figs. 13 and 14B). Boron was detected at a high relative-concentration in one grid well (1.9 percent of the primary aquifers) and at moderate relative-concentrations in 7.4 percent of the primary aquifers (table 4).

Spatially weighted relative-concentrations for three trace elements—aluminum, cadmium, and lead—were high 0.1, 0.3, and 0.9 percent of the primary aquifers, respectively, as compared to 0 percent for these elements for the grid-based approach (table 4). The spatially weighted approach includes data from a larger number of wells than the grid-based approach, and therefore is more likely to include constituents present at high concentrations in small proportions of the primary aquifers.

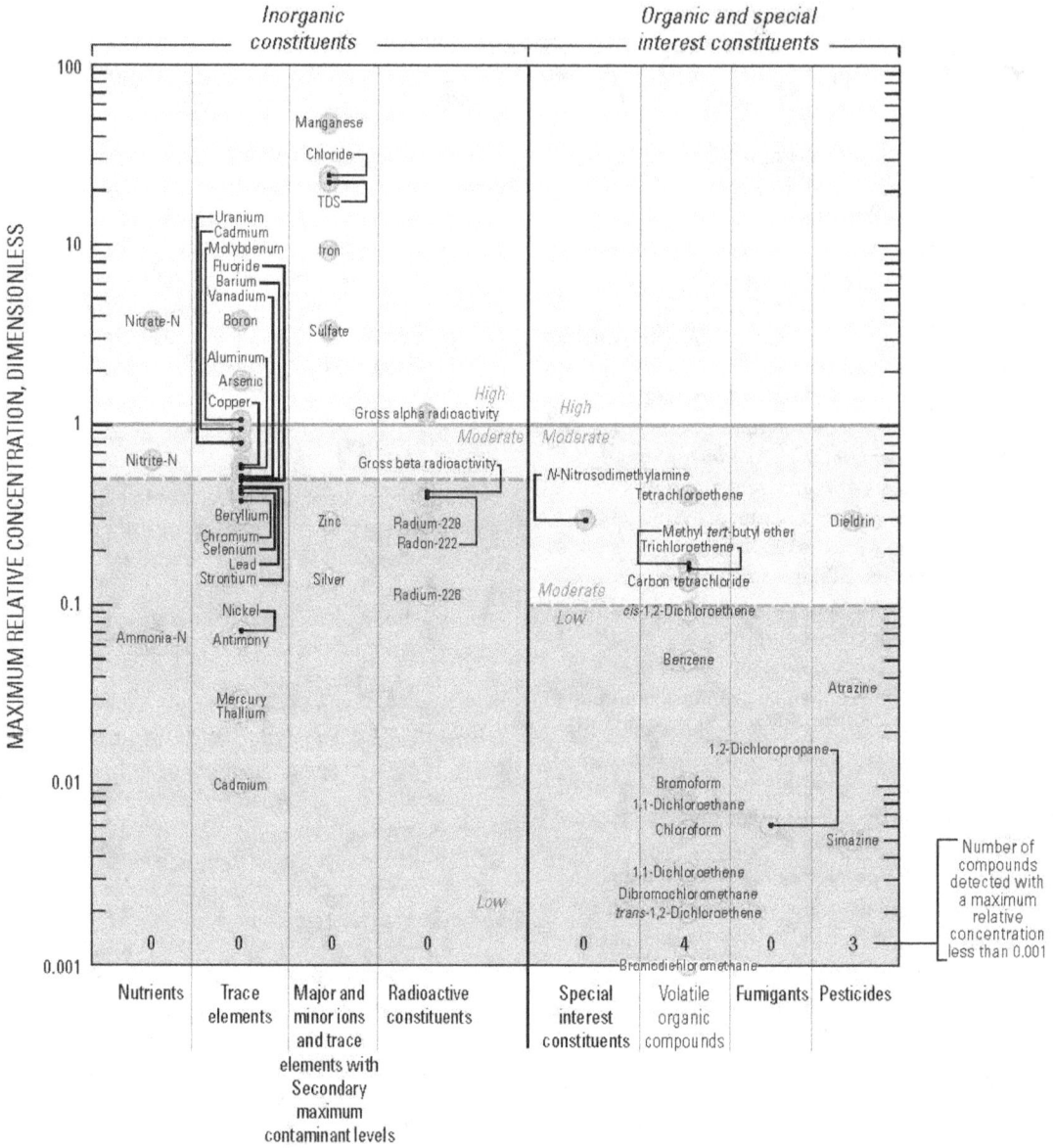

Figure 12.   Maximum relative-concentration of constituents detected in grid wells, by constituent class, Monterey Bay and Salinas Valley Basins study unit, California GAMA Priority Basin Project.

Table 8. Number of constituents analyzed, and, number detected, by the U.S. Geological Survey, with associated benchmarks in each constituent class, Monterey Bay and Salinas Valley Basins study unit, California GAMA Priority Basin Project, July–October 2005.

[Health-based benchmarks include U.S. Environmental Protection Agency (USEPA) and California Department of Public Health (CDPH) maximum contaminant levels (MCL); USEPA lifetime health advisory levels (HAL) and risk-specific dose level at 10^{-5} lifetime cancer risk, and CDPH notification level (NL); RSD5, USEPA risk specific dose at 10^{-5}; AL, USEPA action level; SMCL, USEPA or CDPH secondary maximum contaminant level. VOC, volatile organic compound]

Benchmark type	Organic constituent classes											
	Sum of organic and special interest compounds		VOC and gasoline additives (excluding fumigants)		Fumigants		Pesticides and degradates		Polar pesticides and degradates		Special-interest compounds	
	Number of constituents											
	Analyzed	Detected	Analyzed	Detected	Analyzed	Detected	Analyzed	Detected	Analyzed	Detected	Analyzed	Detected
MCL	46	19	29	16	4	1	3	2	9	0	1	0
HAL	31	4	6	0	1	0	14	3	9	0	1	1
NL	16	3	15	2	0	0	0	0	0	0	1	1
RSD5	7	1	2	0	2	0	3	1	0	0	0	0
AL	0	0	0	0	0	0	0	0	0	0	0	0
SMCL	0	0	0	0	0	0	0	0	0	0	0	0
None	105	5	27	3	2	0	41	2	35	0	0	0
Total	205	32	79	21	9	1	61	8	53	0	3	2

Benchmark type	Inorganic constituent classes										Sum of organic and inorganic constituents	
	Sum of inorganic constituents		Major and minor ions and total dissolved solids		Nutrients		Trace elements		Radioactive constituents			
	Number of constituents											
	Analyzed	Detected	Analyzed	Detected	Analyzed	Detected	Analyzed	Detected	Analyzed	Detected	Analyzed	Detected
MCL	23	23	1	1	2	2	12	12	8	8	69	42
HAL	4	4	0	0	1	1	3	3	0	0	35	8
NL	2	2	0	0	0	0	2	2	0	0	18	5
RSD5	0	0	0	0	0	0	0	0	0	0	7	1
AL	2	2	0	0	0	0	2	2	0	0	2	2
SMCL	6	6	3	3	0	0	3	3	0	0	6	6
None	13	13	7	7	3	3	3	3	0	0	118	18
Total	50	50	11	11	6	6	25	25	8	8	255	82

Organic and inorganic constituents combined	
Analyzed	Detected
261	82

Table 9. Aquifer-scale proportions for constituent classes, Monterey Bay and Salinas Valley Basins study unit, California GAMA Priority Basin Project.

[Aquifer-scale proportions were determined using the grid-based approach unless otherwise noted. Only 29 wells were sampled for *N*-nitrosodimethylamine (NDMA). SMCL, secondary maximum contaminant level]

Constituent class	Constituent not detected (percentage of low aquifer-scale proportion)	Aquifer-scale proportion (percent)		
		Low	Moderate	High
Inorganic constituents with human-health benchmark				
Trace elements		69.0	25.4	5.6
Radioactive constituents		88.1	10.4	1.5
Nutrients		86.8	5.3	7.9
Total for inorganic constituents with human-health benchmarks		50.0	35.5	14.5
Inorganic constituents with aesthetic benchmark				
Total dissolved solids (SMCL)		60.0	31.4	8.6
Major and minor ions (SMCL)		87.1	8.6	4.3
Trace elements (SMCL)		99.7	[1]0.3	0.0
Manganese and (or) iron (SMCL)		67.3	11.3	21.4
Organic constituents with human-health benchmark				
Solvents	87.8	96.6	3.3	[1]0.1
Gasoline additives	92.2	97.7	2.2	[1]0.1
Trihalomethanes	76.5	99.6	0.4	0.0
Other organic compounds	97.8	100.0	0.0	0.0
Herbicides and fumigants	71.4	100.0	0.0	0.0
Insecticides	98.9	98.9	1.1	0.0
Total for organic constituents with human-health benchmarks	48.1	93.2	6.6	[1]0.2
Constituent of special interest				
NDMA; 1,2,3-trichloropropane	89.7	93.1	6.9	0.0

[1]Spatially weighted value.

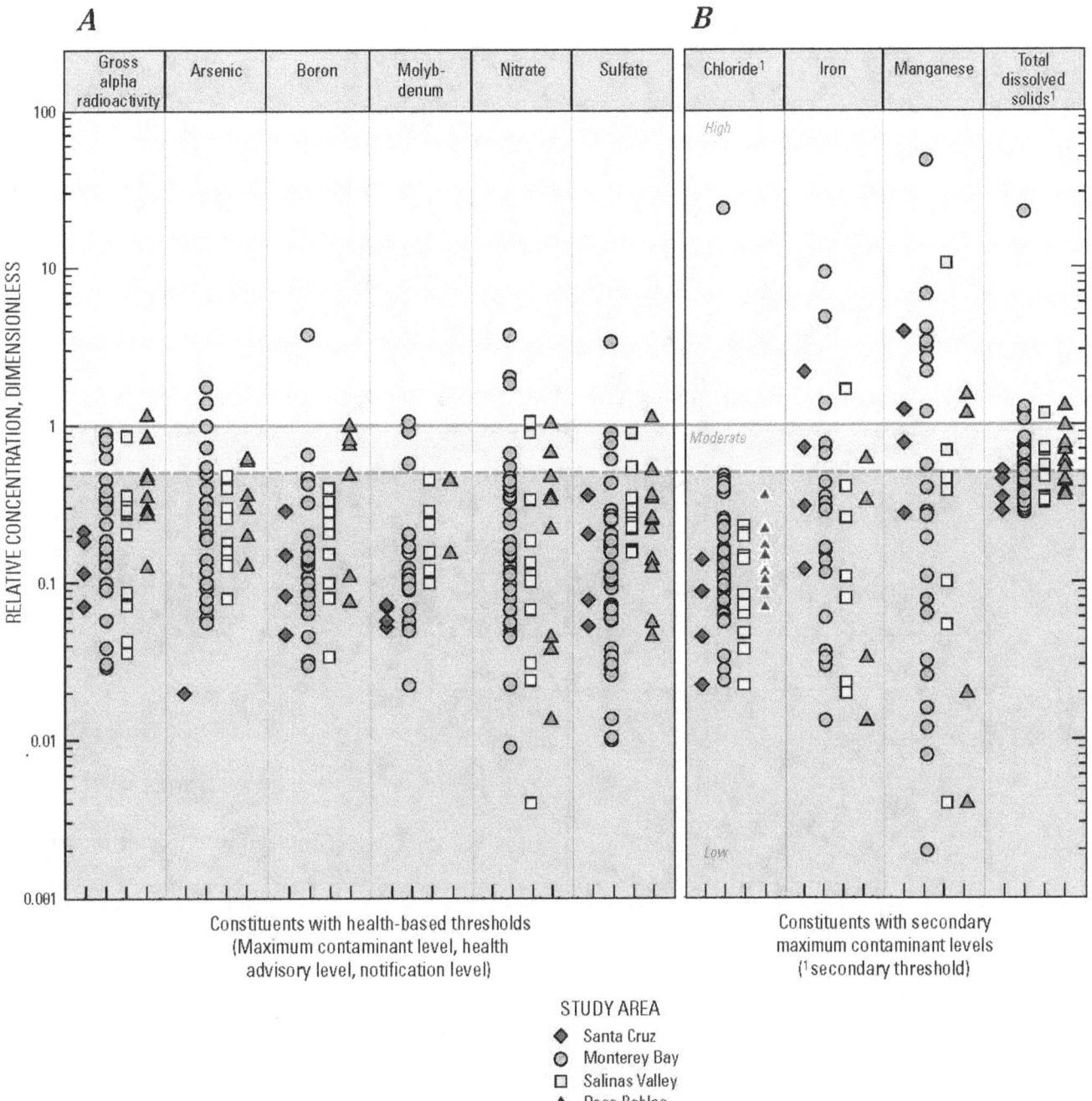

STUDY AREA
◆ Santa Cruz
◉ Monterey Bay
□ Salinas Valley
△ Paso Robles

Figure 13. Relative-concentrations of (*A*) gross alpha radioactivity, arsenic, boron, molybdenum, nitrate, and sulfate with health-based benchmarks, and (*B*) chloride, iron, manganese, and total dissolved solids with aesthetic benchmarks in USGS and CDPH grid wells, Monterey Bay and Salinas Valley Basins study unit, California GAMA Priority Basin Project.

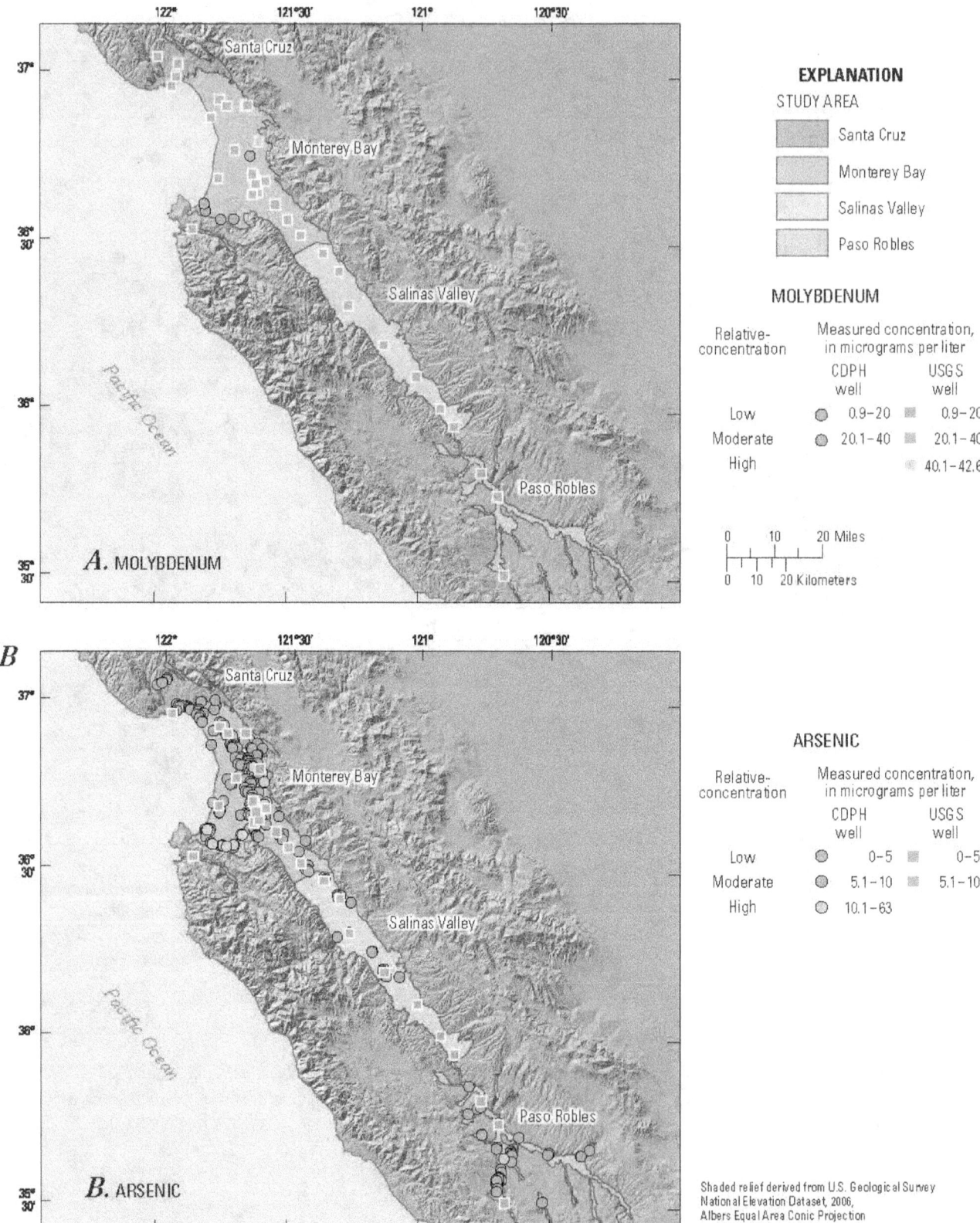

Figure 14. Relative-concentrations of selected inorganic constituents for U.S. Geological Survey USGS-grid and USGS-understanding wells and for the period (July 17, 2002–July 18, 2005) from the California Department of Public Health (CDPH) database), Monterey Bay and Salinas Valley Basins study unit, California GAMA Priority Basin Project.

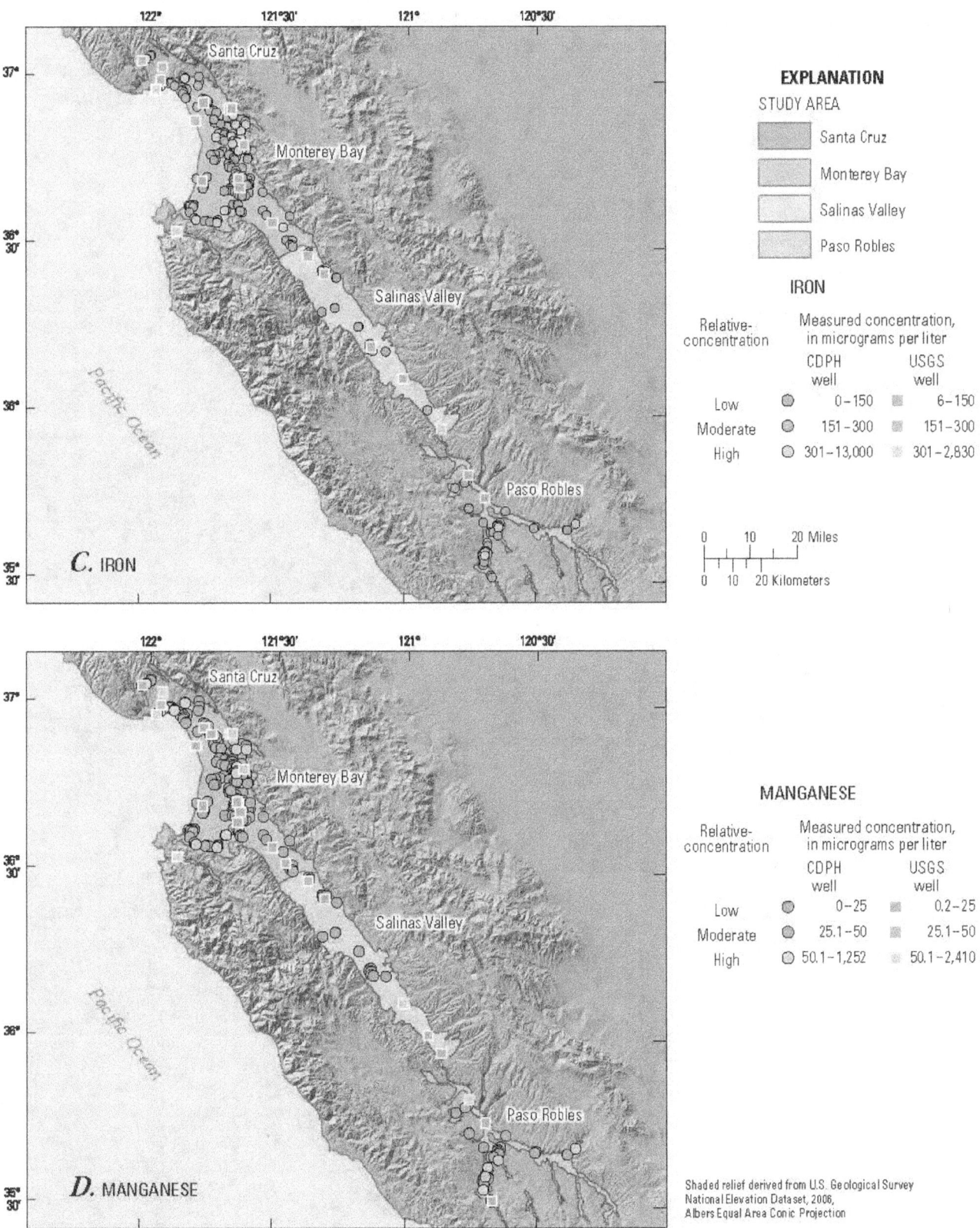

EXPLANATION

STUDY AREA

Santa Cruz

Monterey Bay

Salinas Valley

Paso Robles

IRON

Relative-concentration	Measured concentration, in micrograms per liter	
	CDPH well	USGS well
Low	0–150	6–150
Moderate	151–300	151–300
High	301–13,000	301–2,830

0   10   20 Miles

0   10   20 Kilometers

MANGANESE

Relative-concentration	Measured concentration, in micrograms per liter	
	CDPH well	USGS well
Low	0–25	0.2–25
Moderate	25.1–50	25.1–50
High	50.1–1,252	50.1–2,410

Shaded relief derived from U.S. Geological Survey
National Elevation Dataset, 2006,
Albers Equal Area Conic Projection

Figure 14.—Continued

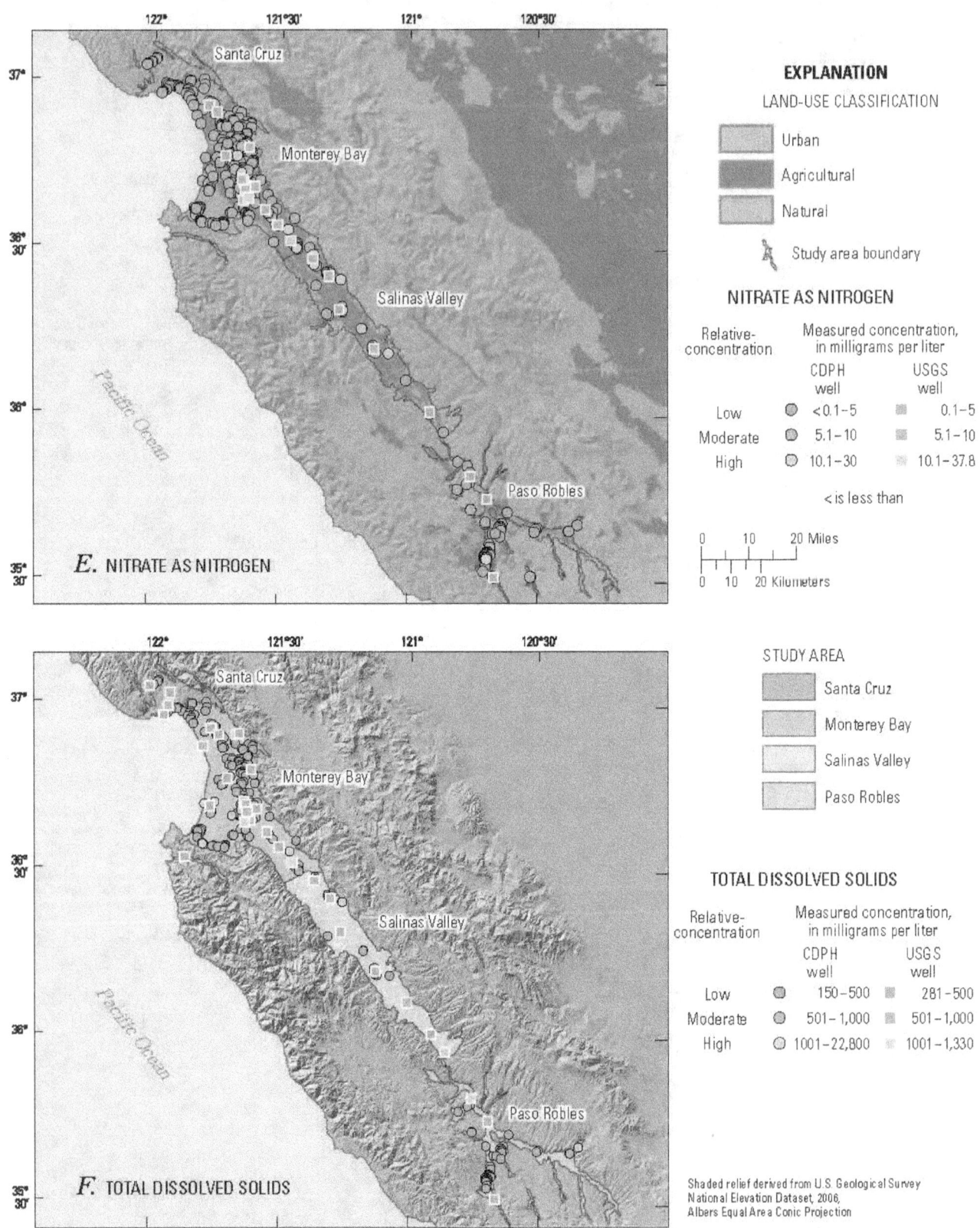

Figure 14.—Continued

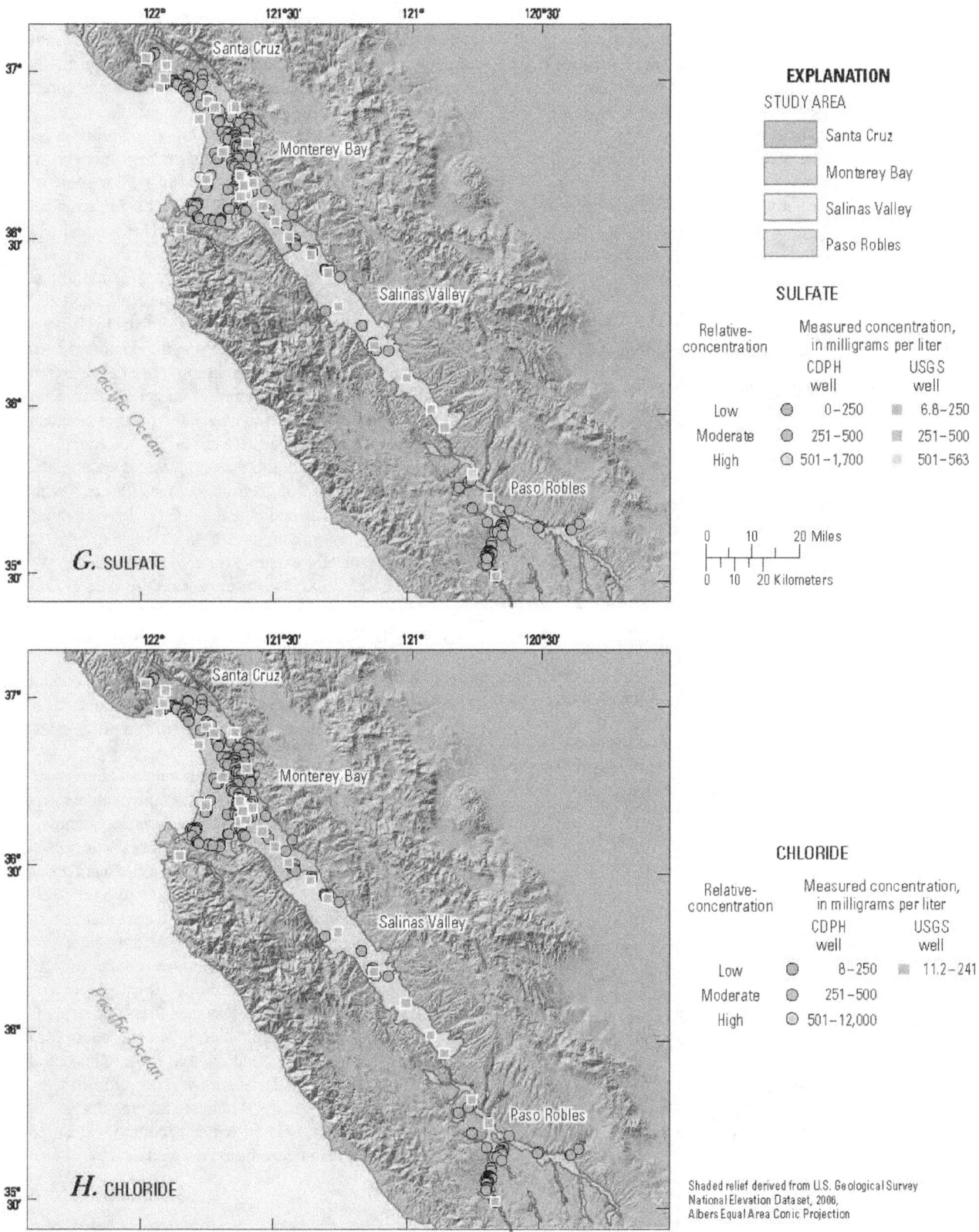

EXPLANATION

STUDY AREA

Santa Cruz

Monterey Bay

Salinas Valley

Paso Robles

SULFATE

Relative-concentration	Measured concentration, in milligrams per liter	
	CDPH well	USGS well
Low	0–250	6.8–250
Moderate	251–500	251–500
High	501–1,700	501–563

0   10   20 Miles

0   10   20 Kilometers

G. SULFATE

CHLORIDE

Relative-concentration	Measured concentration, in milligrams per liter	
	CDPH well	USGS well
Low	8–250	11.2–241
Moderate	251–500	
High	501–12,000	

H. CHLORIDE

Shaded relief derived from U.S. Geological Survey
National Elevation Dataset, 2006,
Albers Equal Area Conic Projection

Figure 14.—Continued

Relative-concentrations for some trace elements—chromium, mercury, vanadium, zinc, and fluoride—were high in at least one well reported in the CDPH database prior to 2002 (table 5), but not during the current period of study (July 17, 2002–July 18, 2005); these high values represent historical values rather than current values.

Among constituents with SMCLs, iron was detected at a high relative-concentration in 7.1 percent of the primary aquifers and moderate relative-concentration in 7.1 percent (table 4, figs. 13 and 14C). Manganese was detected at high relative-concentration in 18.6 percent of the primary aquifers and a moderate relative-concentration in 4.3 percent (fig. 14D).

Understanding Assessment for Molybdenum

Molybdenum is a somewhat rare trace element that is found in the major ore mineral molybdenite, and to a lesser extent, in the mineral wulfenite. Natural sources of molybdenum include low-grade porphyry molybdenum deposits, and as an associated metal sulfide in low-grade porphyry copper deposits. Concentrations measured in water could be directly related to the abundance of molybdenum in mineral species in the environment (Hem, 1970). Industrial uses of molybdenum include steel and iron alloys, ceramics, electrodes, lubricants, adhesive, and catalysts. Molybdenum can accumulate in vegetation and forage crops (particularly legumes) irrigated with water containing molybdenum, or from molybdenum powder used as a fertilizer.

Molybdenum is significantly correlated (positively) with land use classified as natural, as compared with urban land use, and with land use classified as agricultural, as compared with mixed land use (table 7), which may reflect its accumulation in vegetation, or the underlying geology of the region. Relative-concentrations of molybdenum were high and moderate in wells in the Monterey Bay study area (fig. 14A). The HAL-US for molybdenum is 40 µg/L.

Understanding Assessment for Arsenic

Arsenic is a naturally occurring semi-metallic trace element. Potential sources of arsenic to groundwater are both natural and anthropogenic. Natural sources include the dissolution of arsenic-rich minerals, such as arsenian pyrite, a common constituent of shales, and apatite, a common constituent of phosphorites. Arsenic also can be used as a wood preservative, in glass production, in paints, dyes, metals, drugs, soaps, semi-conductors, and in the mining of copper and gold (Welch and others, 2000). Arsenic solubility increases with increasing water temperature, such that hydrothermal fluids often exhibit high arsenic concentrations (Ballantyne and Moore, 1988; Webster and Nordstrom, 2003), as well as in older groundwaters with extended exposure to arsenic-bearing minerals.

Arsenic mobilization and distribution in groundwater is affected by the oxidation-reduction (redox) and pH conditions of the groundwater system (fig. 15). Arsenic is stable in two oxidation states in the environment: arsenite (As^{+3}) and arsenate (As^{+5}). Over a wide pH range and oxic conditions, arsenate (As^{+5}) is predicted to be the predominant species, whereas under more reducing (anoxic) conditions arsenite (As^{+3}) likely would be predominant (Welch and others, 1988). Laboratory reaction experiments by Islam and others (2004) indicate that arsenite was the dominant arsenic species resulting from reductive dissolution of iron oxides by microbial activity and the addition of organic carbon, even though the solid-phase arsenic was in the form of arsenate.

Hydrogen cation concentration (pH) is commonly positively correlated with the concentration of arsenic as a result of the desorption of As from aquifer sediments with pH greater than 7.4 (Belitz and others, 2003; Welch and others, 2006). Previous investigations (Belitz and others, 2003) and reviews of arsenic (for example, Frankenberger, 2002; Welch and others, 2000; 2006; Ravenscroft and others, 2009) have attributed elevated arsenic concentrations in groundwater to two mechanisms: (1) the release of arsenic from dissolution of iron or manganese oxides under iron- or manganese-reducing conditions, and (2) arsenic desorption from aquifer sediments or inhibition of arsenic sorption to aquifer sediments, as a result alkaline groundwater conditions (pH values greater than 8.0).

Evidence for the first mechanism, release of arsenic under reducing conditions, in MS study unit groundwaters includes the association of high and moderate concentrations of arsenic with groundwater having manganese- or iron-reducing conditions. Concentrations of arsenic were greater than 10 µg/L (high relative-concentration) in two grid wells. Concentrations in both of the grid wells were greater than 10 µg/L for arsenic, greater than 100 µg/L for manganese (manganese-reducing conditions), and (or) greater than 100 µg/L for iron (iron-reducing conditions) (fig. 15B). The pH in both grid wells was equal to or less than 7.4, suggesting that reducing conditions rather than high pH may account for the high arsenic concentrations in these two wells. Concentrations of arsenic (5–10 µg/L) and manganese (>50 µg/L) in two additional grid wells (fig. 15B) indicate that reductive dissolution of manganese oxides may account for the moderate concentrations of arsenic in these grid wells.

There is little evidence for the second mechanism, desorption of arsenic from aquifer sediments or inhibition of arsenic sorption to aquifer sediments with increasing pH. pH was equal to or greater than 8 in eight grid wells, and arsenic concentrations in these wells ranged from less than 0.2 to 6.2 µg/L, with a median of 2.2 µg/L. There was no correlation between arsenic concentration and pH (table 10), indicating that the second mechanism—preferential desorption under alkaline conditions—is not the dominant mechanism for high concentrations of arsenic in the MS study unit.

Arsenic distribution was not significantly correlated to redox classification of groundwater in the MS study unit (table 7). This result suggests that several factors, or a combination thereof, are affecting arsenic concentrations in groundwater. Arsenic concentrations in samples with a groundwater age classified as pre-modern were significantly higher than in samples classified as modern or mixed groundwater ages (table 7; fig. 15A). This suggests that groundwater may accumulate arsenic over time from longer exposure to arsenic-bearing minerals.

In summary, data indicate that occurrences of high and moderate arsenic concentrations likely result from the release of arsenic from dissolution of iron or manganese oxides under iron- or manganese-reducing conditions, and accumulation during the relatively long groundwater residence time (groundwater age).

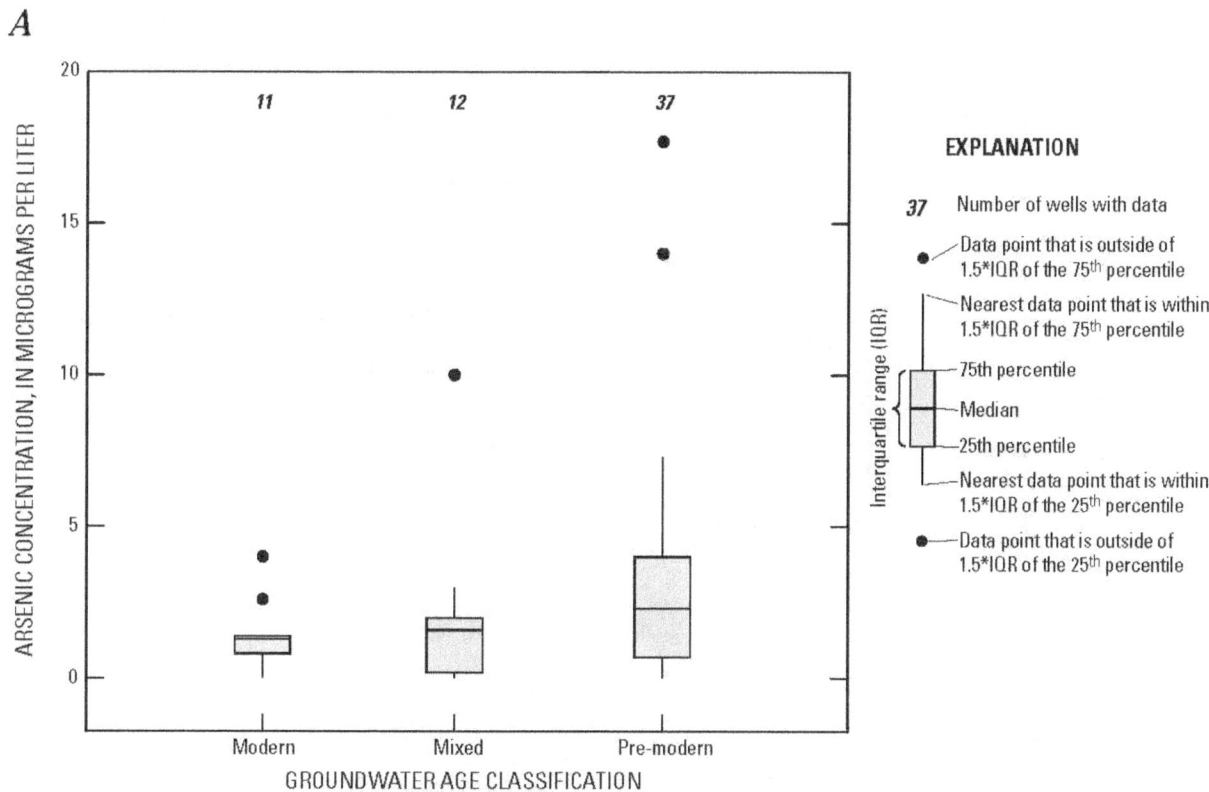

Figure 15. Arsenic concentration relative to (A) classifications of groundwater age, and (B) well depth, manganese and iron concentrations, and pH in grid and understanding wells sampled for the Monterey Bay and Salinas Valley Basins study unit, California GAMA Priority Basin Project.

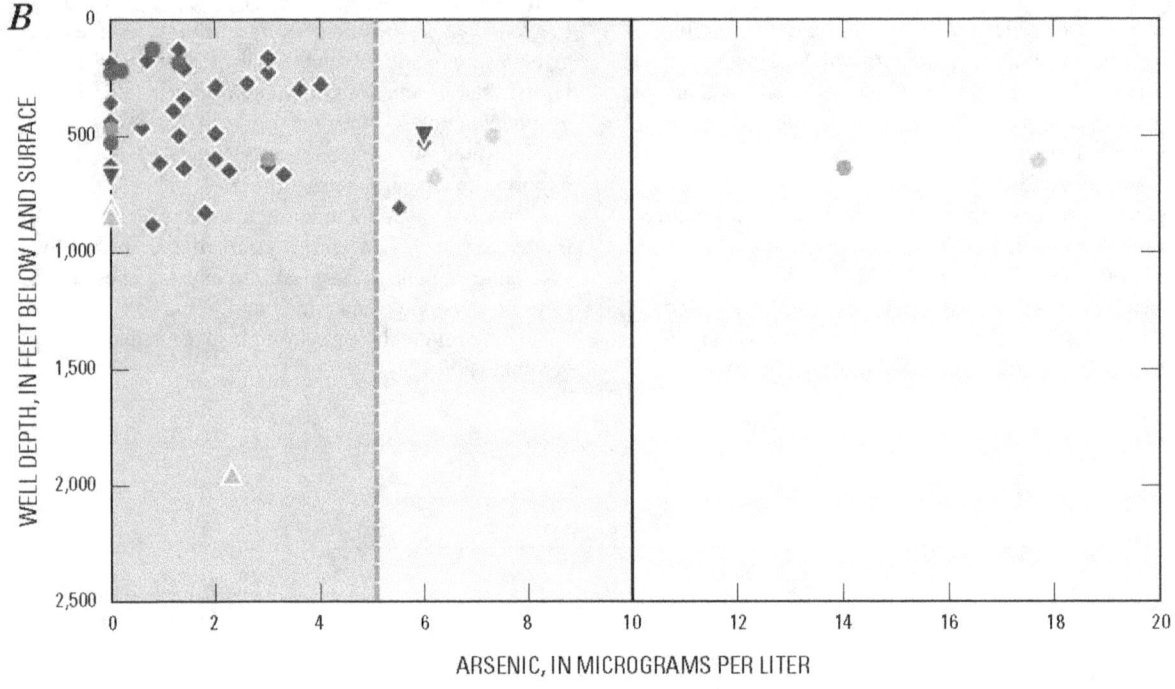

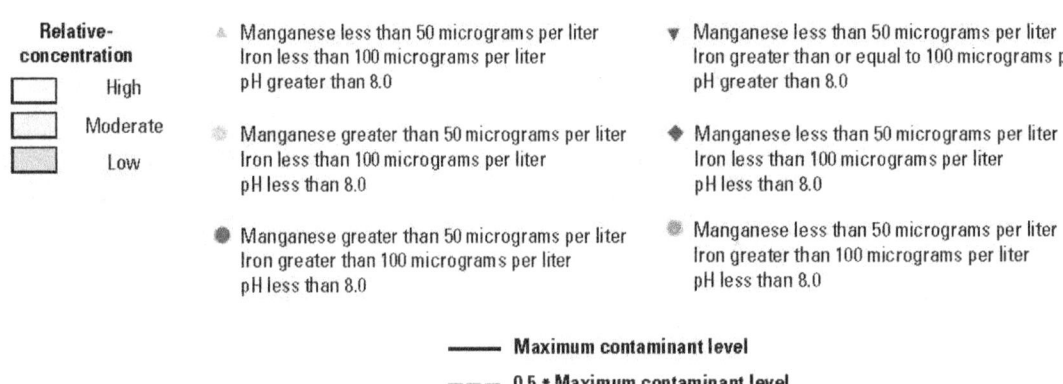

EXPLANATION

Relative-concentration

High
Moderate
Low

△ Manganese less than 50 micrograms per liter
Iron less than 100 micrograms per liter
pH greater than 8.0

● Manganese greater than 50 micrograms per liter
Iron less than 100 micrograms per liter
pH less than 8.0

● Manganese greater than 50 micrograms per liter
Iron greater than 100 micrograms per liter
pH less than 8.0

▼ Manganese less than 50 micrograms per liter
Iron greater than or equal to 100 micrograms per liter
pH greater than 8.0

◆ Manganese less than 50 micrograms per liter
Iron less than 100 micrograms per liter
pH less than 8.0

● Manganese less than 50 micrograms per liter
Iron greater than 100 micrograms per liter
pH less than 8.0

―――― Maximum contaminant level

― ― ― 0.5 * Maximum contaminant level

Figure 15.—Continued

Table 10. Results of non-parametric (Spearman's method) analysis of correlations between selected water-quality constituents and potential explanatory factors, Monterey Bay and Salinas Valley Basins study unit, California GAMA Priority Basin Project.

[Percentage of urban, agricultural, and natural land use measured within a radius of 500 meters around each well included in the analysis. Simazine, single constituents with detection frequency greater than 10 percent of grid wells. Only 29 wells were sampled for NDMA. ρ, Spearman's correlation statistic; PCE, tetrachloroethene; TCE, trichloroethene; NDMA, N-nitrosodimethylamine; MCL-US, USEPA maximum contaminant level; NL-CA, CDPH notification level; HAL-US, USEPA action level; SMCL-CA, CDPH secondary maximum contaminant level; nc, no significant correlation; USEPA, U.S. Environmental Protection Agency; CDPH, California Department of Public Health]

Selected water-quality constituent	Benchmark type	High aquifer-scale proportion, in percent	Grid and understanding wells combined							Grid wells		
			Depth to top-of-perforations	Well depth	Dissolved oxygen	pH	Groundwater temperature	Normalized position of well along flowpath	Altitude of land surface at well	Percentage of urban land use	Percentage of agricultural land use	Percentage of natural land use
			ρ: Spearman's correlation statistic									
Inorganic constituents												
Arsenic	MCL-US	2.8	nc	nc	nc	nc	nc	nc	0.27	nc	nc	nc
Boron	NL-CA	1.9	nc	nc	-0.44	nc	0.33	nc	nc	-0.27	nc	nc
Gross alpha radioactivity	MCL-US	1.5	nc	nc	nc	nc	nc	nc	nc	nc	nc	nc
Molybdenum	HAL-US	2.9	nc	nc	nc	nc	nc	nc	0.53	-0.37	nc	nc
Nitrate (as nitrogen)	MCL-US	7.9	nc	nc	0.45	-0.31	-0.26	nc	nc	nc	0.32	nc
Chloride	SMCL-CA	1.4	nc	nc	nc	nc	0.30	nc	nc	nc	nc	nc
Iron	SMCL-CA	7.1	nc	nc	-0.52	nc	nc	nc	nc	nc	nc	nc
Manganese	SMCL-CA	18.6	nc	nc	-0.47	nc	nc	nc	nc	nc	nc	nc
Sulfate	SMCL-CA	2.9	-0.31	-0.32	-0.28	nc	nc	nc	nc	-0.35	nc	nc
Total dissolved solids (TDS)	SMCL-CA	8.6	nc	nc	-0.26	nc	nc	nc	nc	-0.25	nc	nc
Organic constituents												
Solvents, sum of concentrations (PCE + TCE + carbon tetrachloride)	MCL-US	0.1	nc	nc	nc	nc	nc	0.40	-0.26	0.33	-0.23	nc
Simazine	MCL-US	0.0	-0.36	-0.25	nc	nc	nc	-0.63	0.32	nc	nc	nc
NDMA	NL-CA	6.9	nc	nc	nc	nc	nc	nc	nc	nc	nc	nc

Understanding Assessment for Boron

Although boron was measured at high relative-concentrations in less than 2 percent of the primary aquifers, it is a constituent that affects water quality and is discussed in this section. Boron is a naturally occurring metalloid element that occurs in many minerals. Natural sources of boron include igneous rocks, such as granite and pegmatite (as the mineral tourmaline), and evaporite minerals, such as borax, kernite and colemanite (Hem, 1970; Reimann and Caritat, 1998). Borax, a boron-containing evaporate mineral that is mined in California, is used as a cleaning agent and therefore may be present in sewage and industrial wastes. Seawater contains 4.5 mg/L of boron (Summerhayes and Thorpe, 1996), and boron also is associated with thermal springs (Hem, 1970; Kulongoski and others, 2010). Boron also is used to produce semiconductors, insecticides, preservatives, and chemical reagents. Boron is toxic to plants and humans at high concentrations (Hem, 1970). The NL-CA for boron is 1.0 mg/L.

Boron speciation in groundwater is dependent on pH, salinity, and specific cation content. The neutral form of boron, $B(OH)_3$, is predominant at pH less than 9.2, whereas the anionic form, $B(OH)_4^-$, is predominant at pH greater than 9.2 (Dotsika and others, 2006). Boron is highly mobile because no mineral has a low enough solubility to provide an upper limit to its concentration range.

Boron was detected at high relative-concentrations in 1.9 percent of the primary aquifers. Boron was detected at one grid well in the MB study area at high relative-concentration, and at four grid wells (three PR and one MB) at moderate relative-concentrations. Boron distribution was significantly correlated (negatively) to dissolved oxygen (table 10) and oxic conditions compared with anoxic/suboxic conditions (table 7). This indicates that boron concentrations are higher in groundwaters with anoxic/suboxic than with oxic conditions. Release of boron in anoxic/suboxic conditions could result from the dissolution of oxides on aquifer sediments. The well with the highest boron concentration (3,800 µg/L), MB-DPH-50, also was the well with the highest TDS (22,800 mg/L), chloride (12,000 mg/L), sodium (7,000 mg/L), and sulfate (1,700 mg/L) concentrations. The ratios of the major ions in this well are similar to seawater (Kulongoski and Belitz, 2007), suggesting that seawater intrusion likely is the cause of the high salinity water in this well. However, the dissolution of salts from the saline marine clays that surround the water-bearing zone screened by this well also has been identified as a possible source of the high salinity (Hanson and others, 2002). As mentioned previously, data from this CDPH well were selected to represent a grid cell, and represents poor water quality in the primary aquifers in this location.

Understanding Assessment for Manganese and Iron

Potential natural sources of manganese and iron in groundwater include the dissolution of igneous and metamorphic rocks as well as the dissolution of various secondary minerals (Hem, 1970). Some rocks that contain significant amounts of manganese and iron have a relatively high composition of the minerals olivine, pyroxene, and amphibole. Potential anthropogenic sources of iron and manganese in groundwater include effluents associated with the steel and mining industries (Reimann and Caritat, 1998), and soil amendments in the form of manganese and iron sulfates that are added to deficient soils in order to stimulant crop growth. Distributions of iron and manganese concentrations are strongly influenced by redox conditions in the aquifer. In sediments, the oxyhydroxides of manganese and iron are common as coatings on mineral surfaces and as suspended particles (Sparks, 1995). These oxyhydroxides are stable in oxygenated systems at a neutral pH. However, under anoxic conditions, the process of reductive dissolution releases these minerals, which affect the mobility of manganese and iron in aquifer systems (Sparks, 1995).

In the MS study unit, concentrations of manganese and iron were significantly correlated (negatively) with oxic compared with mixed conditions, and oxic compared with anoxic/suboxic conditions (table 7). Both iron and manganese were significantly correlated (negatively) to dissolved oxygen (table 10), indicating that reductive dissolution is a significant pathway for the mobilization of manganese and iron in groundwater in the MS study unit. Relative-concentrations of iron and manganese were high and moderate in the SC, MB, SV, and PR study areas (figs. 13, 14C, and 14D) reflecting the natural distribution of iron- and manganese-reducing conditions that result from reductive dissolution of iron and manganese oxides in the aquifer sediments.

Radioactive Constituents

The high relative-concentrations of radioactive constituents was 1.5 percent in the MS study unit (table 9), reflecting the detection of gross alpha radioactivity (72-hour count). Gross alpha radioactivity was detected at high relative-concentrations in 1.5 percent of grid wells, and at moderate relative-concentrations in 10.4 percent (table 4). In addition, radium (combined radium-228 and radium-226) was detected at high relative-concentrations during July 17, 2002–July 18, 2005; however, these high values were not the most recent values from the CDPH database representing those wells (table 4).

Nutrients

Nutrients as a class was detected at high relative-concentrations in 7.9 percent of the primary aquifers and moderate in 5.3 percent (table 9), resulting from the detection of nitrate plus nitrite, as nitrogen (hereinafter referred to as nitrate) (table 4). Nitrate was detected at high relative-concentrations in 7.9 percent of grid wells, and at moderate relative-concentration in 5.3 percent of the grid wells (table 4; fig. 14E). Nitrite was detected at high relative-concentrations in 0 percent of grid wells, and at moderate relative-concentration in 1.4 percent of the grid wells (table 4).

Understanding Assessment for Nitrate

Nitrogen in groundwater occurs in the forms of dissolved nitrate, nitrite, and ammonia. Certain bacteria and algae naturally convert nitrogen from the atmosphere to nitrate, which is an important nutrient for plants. Nitrate also is present in trace amounts in precipitation and is produced by desert plants (Hem, 1970). Anthropogenic sources of nitrate include application as a fertilizer for agriculture; and livestock, when in concentrated numbers, produce nitrogenous waste that can leach into groundwater. Septic systems also may introduce nitrogenous waste into groundwater. In addition, nitrate may be associated with uranium mining and processing (Hem, 1970).

Nitrate concentrations were slightly higher in wells with groundwater ages classified as modern or mixed than in wells with groundwater ages classified as pre-modern (fig. 16A). Nitrate concentrations in wells classified as natural land use had significant negative correlation with agricultural land use (table 7 and fig. 14E), and nitrate concentrations in wells classified as urban/agricultural land use were higher than in wells classified as natural land use (fig. 16B). Nitrate concentrations had significant positive correlation with dissolved oxygen and percentage of agricultural land use, and had negative correlation with pH and groundwater temperature (table 10). The positive correlation between nitrate and agricultural land use (table 10) suggests that the nitrate likely is from agricultural sources.

Relative-concentrations of nitrate in three USGS-grid wells and one USGS-understanding well were high (greater than 10 mg/L as nitrogen) (fig. 16B). These wells had an urban/agricultural land-use classification, less than 350 ft to the tops of the well perforations, and mixed or modern age classifications (fig. 16B). Four additional wells had moderate relative-concentrations of nitrate, two of which had urban/agricultural land-use classification and two of which had mixed land-use classification. All four moderate relative-concentration samples were from wells with the top of the well perforations less than 200 ft, and groundwater age in three of the wells was classified as modern or mixed; groundwater age in one of the wells was classified as pre-modern.

Major and Minor Ions

The major ions chloride and sulfate, and TDS have upper SMCL-CA benchmarks that are based on aesthetic properties. The minor ion fluoride has an MCL-US, and the remaining seven major or minor ions do not have benchmarks (table 8).

TDS was detected at a high relative-concentration in 8.6 percent and a moderate relative-concentration in 31.4 percent (table 4; fig. 14F). Sulfate was detected at a high relative-concentration in 2.9 percent of the primary aquifers and a moderate relative-concentration in 8.6 percent (table 4; fig. 14G). Chloride was detected at a high relative-concentration in 1.4 percent of the primary aquifers (table 4; fig. 14H). Relative-concentrations of chloride, sulfate, and TDS were high in a single well in the CDPH database, which represents poor water quality in the primary aquifers in this location.

Understanding Assessment for Total Dissolved Solids

Natural sources of TDS include seawater intrusion, mixing of groundwater with deep saline groundwater (connate water) that is influenced by interactions with deep marine or lacustrine sediments, concentration of salts by evaporation in discharge areas, and (or) water-rock interactions. Potential anthropogenic sources of TDS to groundwater in the MS study unit include evaporation from agricultural and urban irrigation, disposal of wastewater and industrial effluent, and leaking water and sewer pipes. The anion chloride is a major component of TDS, and its distribution, for the most part, reflects that of TDS.

In the MS study unit, TDS had a significant negative correlation with dissolved oxygen and percentage of urban land use (table 10). TDS had a significant positive correlation with natural compared with mixed land use, agricultural compared with mixed land use, and urban compared with mixed land use (table 7). Figure 17 compares the TDS and the altitude of land surface (above sea level) for MS study unit wells, as a function of the normalized position of the wells along the flowpath (proximal, medial, or distal). Concentrations of TDS in proximal flowpath wells generally were higher than in medial wells in the SV study area. In the MB study area, concentrations of TDS were higher in medial wells than in distal flowpath wells. The decrease of TDS concentrations in groundwater along the flowpath in the MB study area may be explained by the dilution of groundwater with low-TDS reservoir water used for irrigation. Relative-concentrations of TDS were high and moderate in wells in all study areas (fig. 14F). High concentrations of TDS (and Cl and SO_4) near Monterey Bay may result from seawater intrusion; however, numerous thick fine-grained interbeds and confining units in the aquifer systems (fig. 10) limit the vertical movement of fresh and saline groundwater between aquifers. High TDS groundwater was observed at depths from 234 ft to 800 ft in the MB study area.

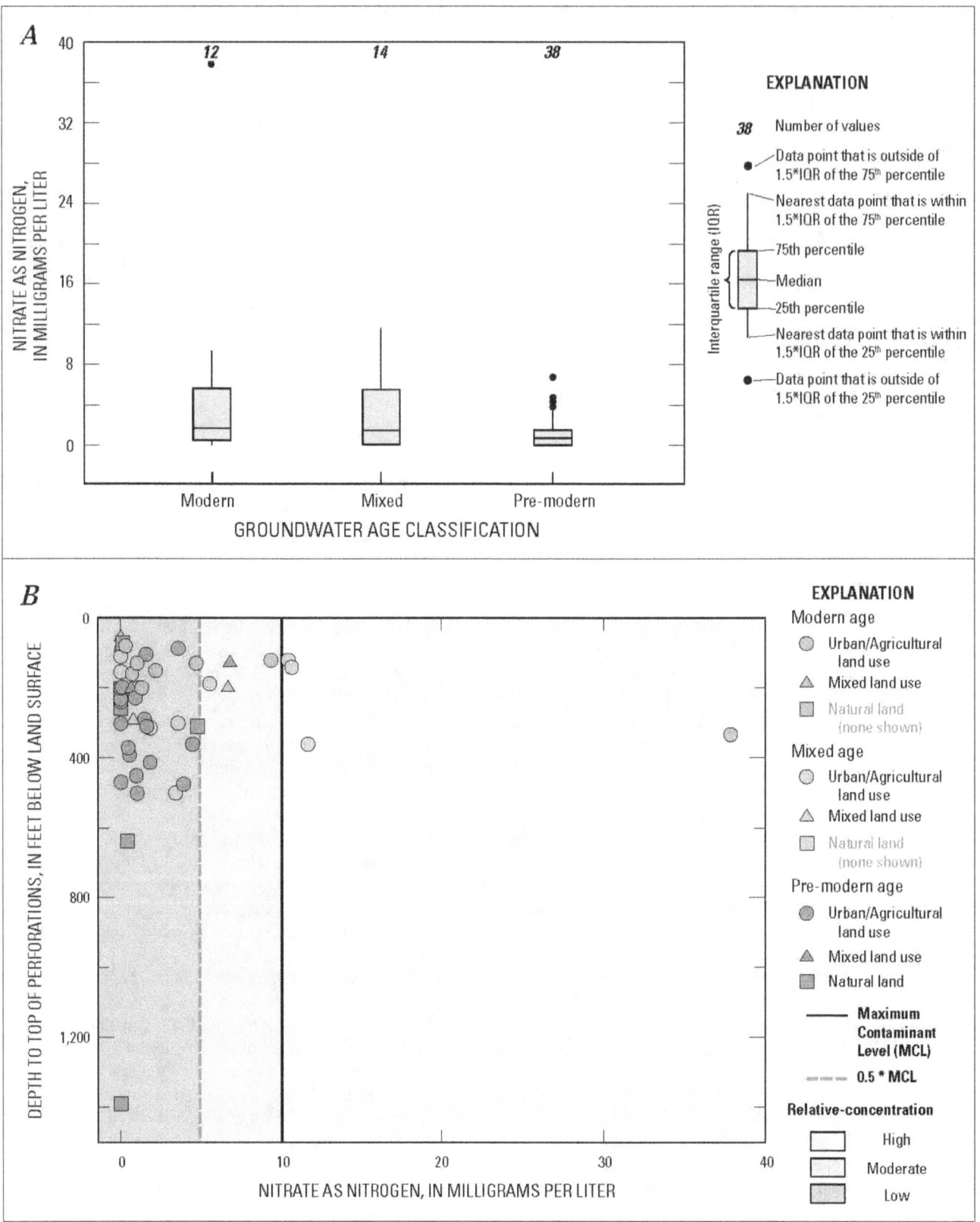

Figure 16.   Nitrate, as nitrogen, concentrations relative to (*A*) classifications of groundwater age, and (*B*) depth to top of perforations, classification of groundwater age, and land use, in USGS-grid and USGS-understanding wells sampled for the Monterey Bay and Salinas Valley Basins study unit, California GAMA Priority Basin Project.

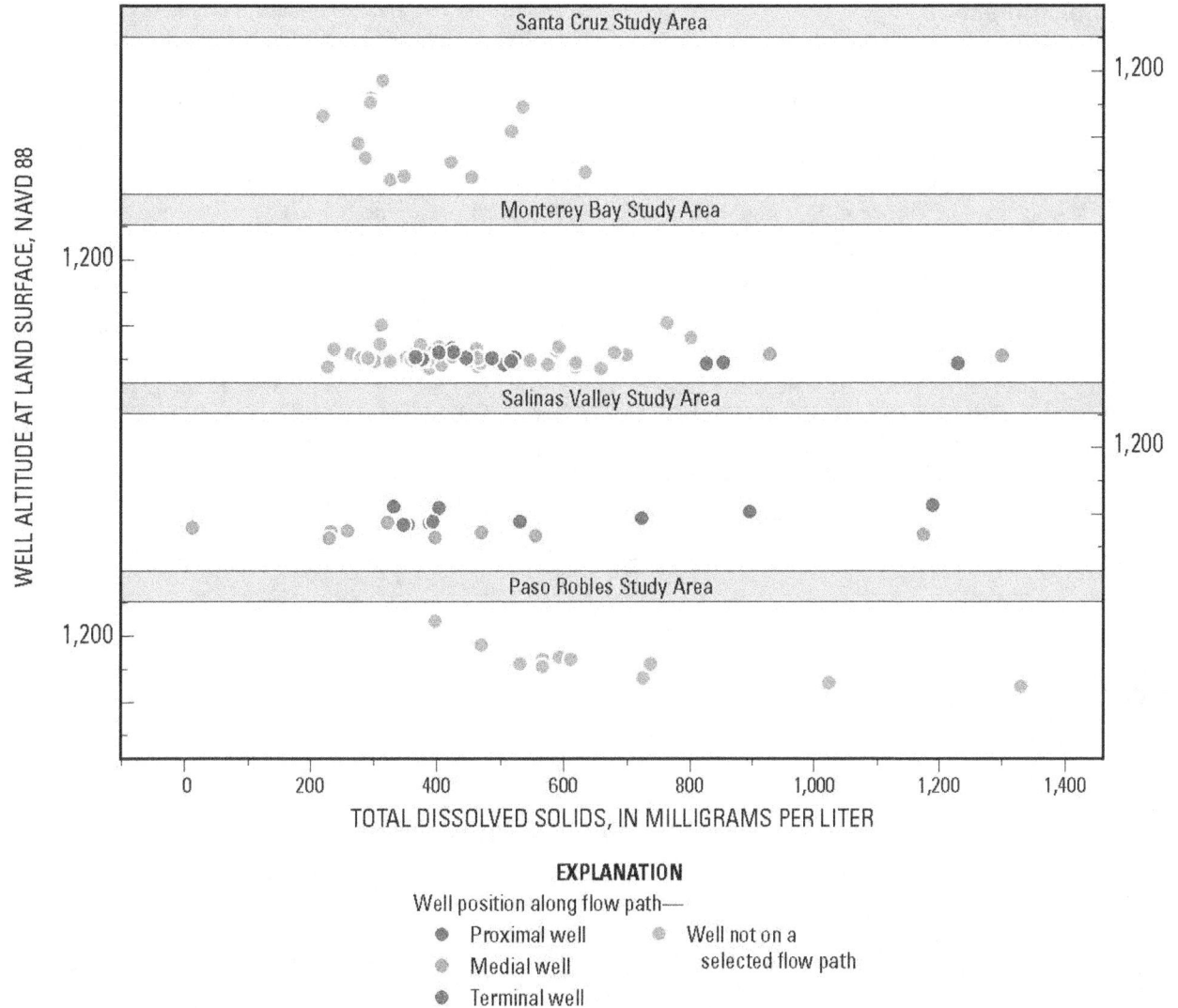

Figure 17. Well altitude at land surface, and total dissolved solid concentrations, in grid and understanding wells in the Santa Cruz, Monterey Bay, Salinas Valley, and Paso Robles study areas of the Monterey Bay and Salinas Valley Basins study unit, California GAMA Priority Basin Project. Positions along the flowpath (proximal, medial, or distal) are indicated for some wells in the Salinas Valley flowpath.

Understanding Assessment for Sulfate

Natural sources of sulfate include the dissolution of natural sulfur and its oxidation to the anion sulfate, or the biochemical oxidation of sulfide minerals or species. Sulfate occurs in evaporite sediments as gypsum or anhydrite, and is common in rainfall (commonly exceeding 1 mg/L) (Hem, 1970). The sulfate in rainfall has been attributed to the emission of H_2S at the ocean margins, the combustion of fuels, emissions from volcanoes, springs, and fumaroles, the solution of dust particles, dissolution of gypsum or anhydrite, and the oxidation of uplifted fine-grained marine sediments (Hem, 1970).

In the MS study unit, sulfate had a significant negative correlation with depth to the tops of the perforations, well depth, dissolved oxygen, and percentage of urban land use (table 10). Sulfate also had a significant positive correlation with natural compared with mixed land use, agricultural compared with mixed land use, and urban compared with mixed land use (table 7). The land-use correlations are inconclusive, but the negative correlations of sulfate with well depth and depth to top of perforations suggest that sulfate concentrations are higher in shallow groundwater than in deep groundwater. Relative-concentrations of sulfate were high and moderate in wells in the PR, SV, and MB study areas (fig. 14G).

Organic Constituents

The organic compounds are organized by constituent class, including four classes of volatile organic compounds (VOCs) and two classes of pesticides. VOCs may be in paints, solvents, fuels, and refrigerants; VOCs can be byproducts of water disinfection and are characterized by a volatile nature, or tendency to evaporate. In this report, VOCs are classified into four categories: (1) solvents, (2) gasoline additives, (3) trihalomethanes, and (4) other organic compounds (including organic synthesis reagents and refrigerants). Pesticides are used to control weeds, fungi, or insects in agricultural, urban, and suburban settings. In this report, pesticides are grouped into two classifications: herbicides and fumigants; or, insecticides. Organic constituents were detected in 45.1 percent of the 91 USGS-grid wells in the MS study unit. Thirty-two of the 205 organic compounds analyzed

for were detected, and human-health benchmarks (table 8) have been established for most (27 of 32) of these organic constituents.

The proportion of the aquifer with high relative-concentrations of organic constituents was 0.2 percent (table 9), based on the spatially-weighted approach. The solvent PCE (0.1 percent) and the gasoline oxygenate MTBE (0.1 percent) were detected at high relative-concentrations (table 4). The proportion of the aquifer with moderate relative-concentrations of organic constituents was 6.6 percent (table 9). Tetrachloroethene (PCE), dieldrin, methyl *tert*-butyl ether (MTBE), trichloroethene (TCE), and carbon tetrachloride were detected at moderate relative-concentrations (figs. 12, 18, 19). Simazine was detected in more than 10 percent of the primary aquifers (fig. 19A-19B).

The constituents of special interest NDMA and 1,2,3-TCP were detected at a low concentration in two wells and in one well, respectively, in the MS study unit.

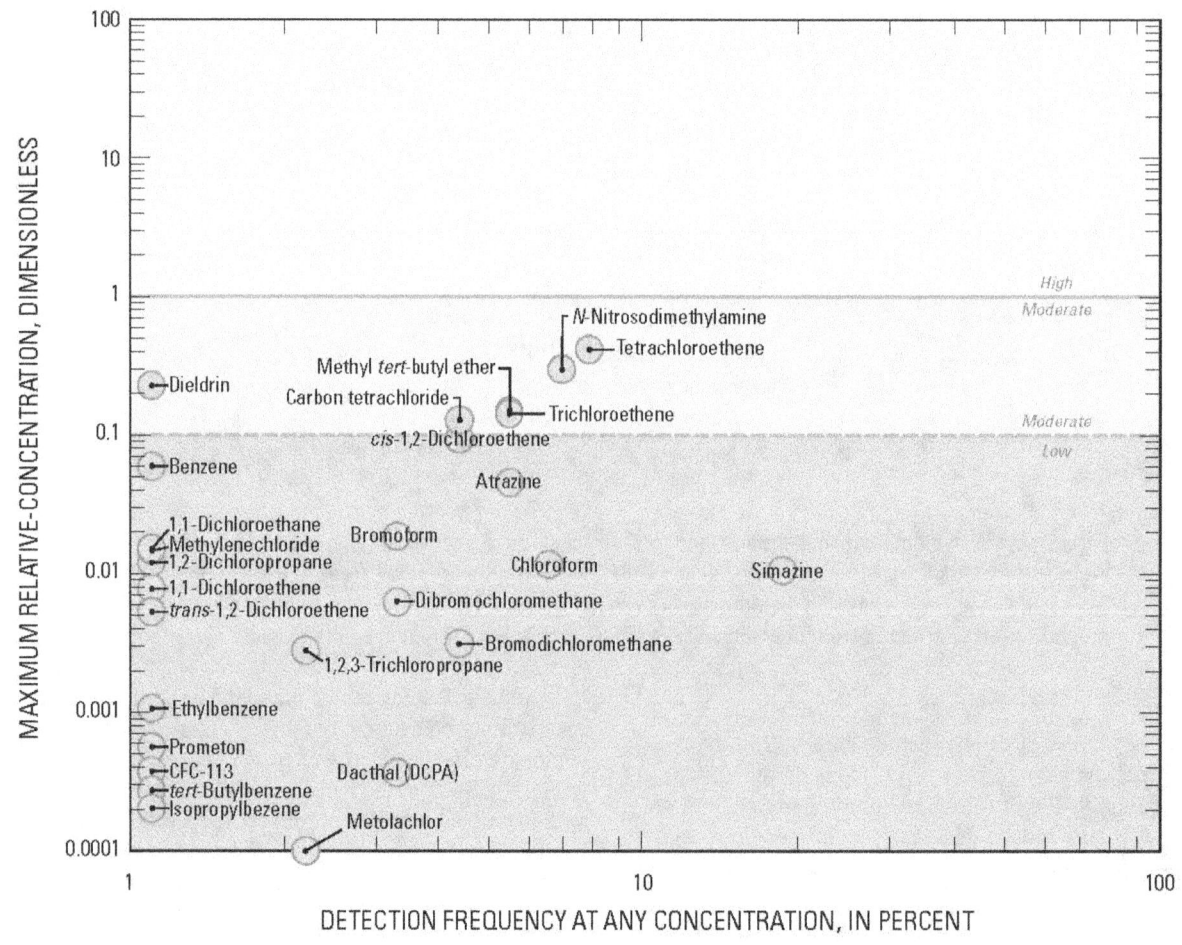

EXPLANATION

Atrazine **Name and center of symbol is the maximum relative concentration for that constituent—**
Unless indicated by following location line:

Figure 18. Detection frequency and maximum relative-concentration of organic and special-interest constituents detected in USGS-grid wells in the Monterey Bay and Salinas Valley Basins study unit, California GAMA Priority Basin Project.

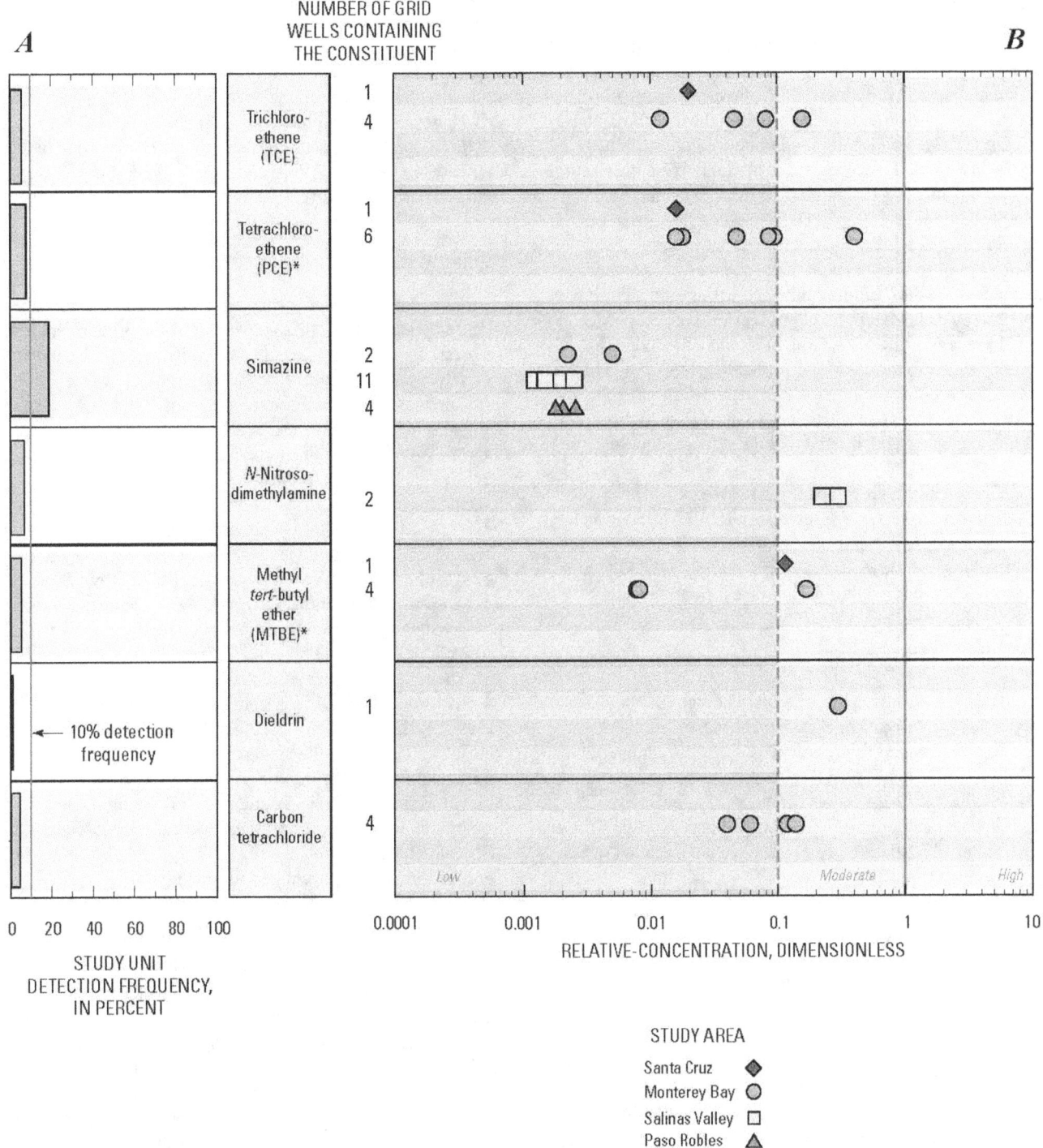

Figure 19.   (*A*) Detection frequency and (*B*) relative-concentrations of selected organic and special-interest constituents in USGS-grid wells in the Monterey Bay and Salinas Valley Basins study unit, California GAMA Priority Basin Project, July–October 2005. *Constituent with spatially-weighted high aquifer-scale proportion.

Solvents

Solvents are used for various industrial, commercial, and domestic purposes. The solvent PCE had a spatially-weighted high aquifer-scale proportion of 0.1 percent (table 4), and the solvent 1,4-dioxane was recorded in the CDPH database at a high relative-concentration in one well during July 17, 2002–July 18, 2005; however, the high relative-concentration was not the most recent value from the CDPH data used to represent that well (table 4). PCE primarily is used for dry-cleaning of fabrics and degreasing metal parts, and is an ingredient in a wide range of products including paint removers, polishes, printing inks, lubricants, and adhesives. Solvents as a class were at a high aquifer-scale proportion of 0.1 percent of the primary aquifer, and a moderate aquifer-scale proportion of 3.3 percent (table 9). None of the individual solvent compounds were detected in more than 10 percent of the wells tested, nor were relative-concentrations high in greater than 0.1 percent of the primary aquifers (table 4).

Historically high values for the solvents 1,1-dichloroethane, dichloromethane, 1,2-dichloroethane, 1,1,2,2-tetrachloroethane, and 1,2,4-trichlorobenzene were recorded in the CDPH database prior to July 17, 2002 (table 5) but were not recorded during the current period of study.

Gasoline Additives

Gasoline additives were detected at high (0.1 percent) and moderate (2.2 percent) relative-concentrations in the primary aquifers, as a result of the detection of the discontinued gasoline oxygenate MTBE (table 4).

Historically high values for the VOCs benzene, naphthalene, and toluene were recorded in the CDPH database prior to July 17, 2002 (table 5) but were not recorded during the current period of study.

Trihalomethanes

The category "trihalomethanes" was detected at high (0 percent) and moderate (0.4 percent) relative-concentrations in the primary aquifers (table 4).

The constituents bromodichloromethane and chloroform were detected at high relative-concentrations in the CDPH database during July 17, 2002–July 18, 2005, but these high relative-concentrations were not the most recent value selected for calculating aquifer-scale proportion (table 4).

Other Organic Compounds

Other organic compounds, includes organic synthesis reagents and refrigerants; there were no grid-based high or moderate relative-concentrations (table 4).

Historically high values for 1-dichloroethene, di(2-ethylhexyl)phthalate, and vinyl chloride were recorded in the CDPH database prior to July 17, 2002 (table 5) but were not recorded during the current period of study.

Herbicides and Fumigants

As a class, herbicides and fumigants were detected at low relative-concentrations in 28.6 percent of the primary aquifers, however, they were not detected at moderate or high relative-concentrations. Low relative-concentrations of the herbicide simazine were detected in samples from the MS study unit (figs. 18 and 19). Simazine was detected in 18.7 percent of the grid wells (fig. 19); the maximum relative-concentration was 0.005 µg/L. Simazine was among the most commonly detected herbicides in groundwater in major aquifers across the United States (Gilliom and others, 2006). Historically, simazine most commonly is used on vineyards and orchards in the MS study unit but also is used on rights-of-way for weed control (Domagalski and Dubrovsky, 1991). Simazine was the most frequently detected triazine herbicide in groundwater in California (Troiano and others, 2001).

The relative-concentrations for the fumigant 1,4-dichlorobenzene were high in the CDPH database during July 17, 2002–July 18, 2005, but this high relative-concentration was not the most recent value selected for calculating aquifer proportion (table 4). Historically high values for the herbicide atrazine, and the fumigants bromomethane and ethylene dibromide were recorded in the CDPH database prior to July 17, 2002 (table 5), but were not recorded during the current period of study.

Understanding Assessment for Simazine

Simazine was detected in 17 grid wells, 14 with top perforations less than 200 ft below land surface. Simazine had a negative correlation with the depth to the top of perforation, well depth, normalized position along the flowpath, and a positive correlation with altitude of land surface at the well (table 10). Simazine concentrations also had a negative correlation with natural compared with agricultural land-use classification (table 7). Seventeen grid wells with simazine detections were classified as agricultural (12 wells), urban (4 wells), or natural (1 well). Simazine has been used in agricultural applications on citrus and vineyards, and in urban settings for weed control (Gilliom and others, 2006). Most of the wells in which simazine was detected may be characterized as shallow (less than 200 ft) and in agricultural or urban land-use areas (fig. 20).

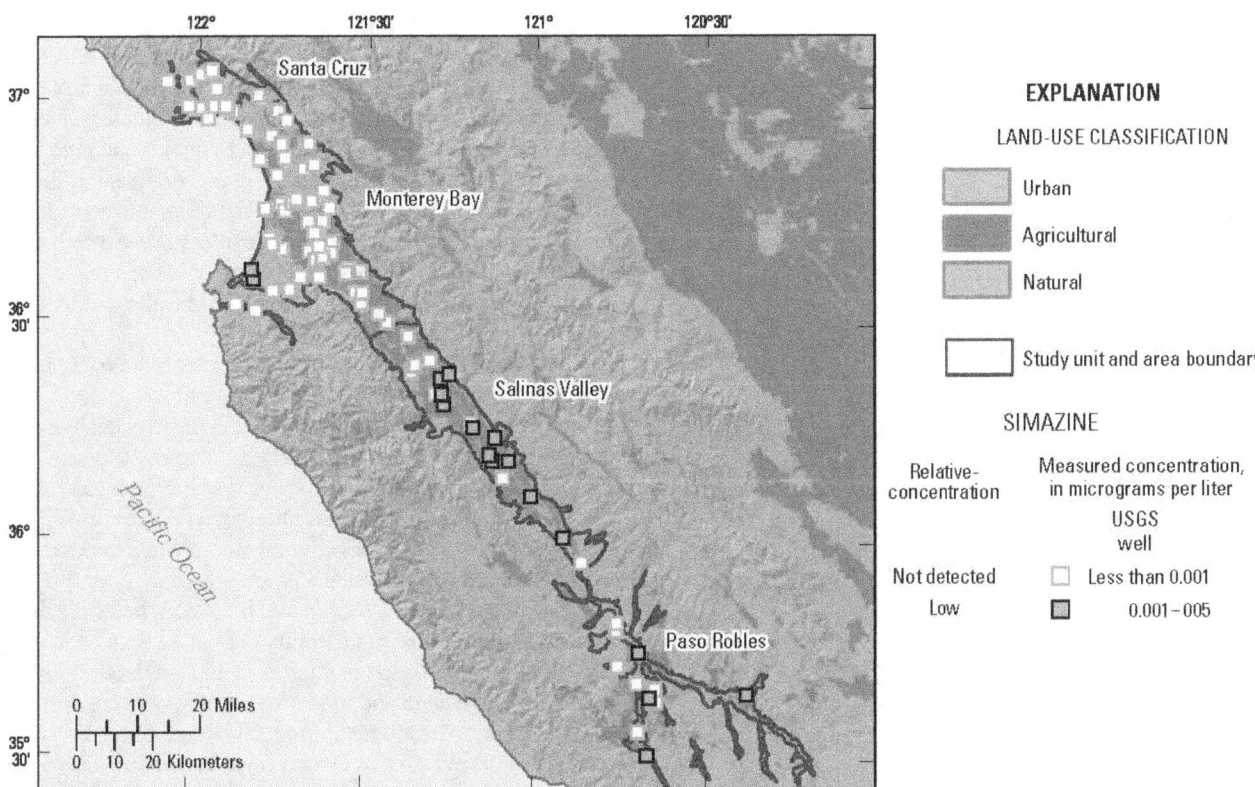

Figure 20.   Land-use classifications and relative-concentrations of the herbicide simazine in U.S. Geological Survey (USGS)-grid wells (2002–2005), and for the period (July 17, 2002–July 18, 2005) from the California Department of Public Health (CDPH) database, Monterey Bay and Salinas Valley Basins study unit, California GAMA Priority Basin Project.

Insecticides

The insecticide dieldrin was detected at one grid well at a moderate relative-concentration (figs. 18 and 19). Historically high values for the insecticide heptachlor were recorded in the CDPH database for the period before July 17, 2002 (table 5) but not during the current period of study.

Special-Interest Constituents

Constituents of special interest analyzed for the MS study unit were NDMA, 1,2,3-TCP, and perchlorate. These constituents were selected because they recently have been detected in drinking-water supplies, or are considered to have the potential to reach drinking-water supplies (California Department of Public Health, 2008a, 2008b, and 2008c). NDMA was detected in two wells (out of 29 wells sampled) at moderate relative-concentrations (table 4). The NDMA data units were incorrectly reported in the Data Series Report as micrograms per liter (μg/L); the units in the data report should have been nanogram per liter (ng/L). 1,2,3-TCP was detected in one well at low relative-concentration. Perchlorate was not detected in the 29 grid wells sampled (Kulongoski and Belitz, 2007).

Summary

Groundwater quality in the approximately 1,000-square-mile (2,590 km²) Monterey Bay and Salinas Valley Basins (MS) study unit was investigated as part of the Priority Basin Project of the Groundwater Ambient Monitoring and Assessment (GAMA) Program. The GAMA MS study provides a spatially unbiased characterization of untreated groundwater quality in the primary aquifers. The assessment is based on water-quality and ancillary data collected in 2005 by the U.S. Geological Survey (USGS) from 97 wells, and on water-quality data from the California Department of Public Health (CDPH) database.

The first component of this study, the status of the current quality of the groundwater resource, was assessed by using data from samples analyzed for volatile organic compounds (VOCs), pesticides, and naturally occurring inorganic constituents, such as major ions and trace elements. The *status assessment* characterizes the quality of groundwater resources in the primary aquifers of the MS study unit, not the treated drinking water delivered to consumers by water purveyors.

Relative-concentrations (sample concentration divided by the health- or aesthetic-based benchmark concentration) were used for evaluating groundwater quality for those constituents that have Federal and (or) California regulatory or non regulatory benchmarks for drinking-water quality.

Aquifer-scale proportion was used as the primary metric for evaluating regional-scale groundwater quality. High aquifer-scale proportion is defined as the percentage of the primary aquifers with relative-concentration greater than 1.0 for a particular constituent or class of constituents; proportion is based on an areal rather than a volumetric basis. Moderate and low aquifer-scale proportions were defined as the percentage of the primary aquifers with moderate and low relative-concentrations, respectively. Two statistical approaches, grid-based and spatially weighted, were used to evaluate aquifer-scale proportions for individual constituents and classes of constituents. Grid-based and spatially weighted estimates were comparable in the MS study unit (within 90-percent confidence intervals). However, the spatially weighted approach was superior to the grid-based proportion when relative-concentrations of a constituent are high in a small fraction of the aquifer.

Inorganic constituents with human-health benchmarks were detected at high relative-concentrations in 14.5 percent of the primary aquifers, moderate in 35.5 percent, and low or not detected in 50.0 percent. The high aquifer-scale proportion of inorganic constituents primarily reflected high aquifer-scale proportions of nitrate (7.9 percent), molybdenum (2.9 percent), arsenic (2.8 percent), boron (1.9 percent), and gross alpha radioactivity (1.5 percent). Relative-concentrations of organic constituents (one or more) were high in 0.2 percent, moderate in 6.6 percent, and low in 93.2 percent (not detected in 48.1 percent) of the primary aquifers. The high aquifer-scale proportion of organic constituents primarily reflected high aquifer-scale proportions of PCE (0.1 percent) and MTBE (0.1 percent). The inorganic constituents with secondary maximum contaminant levels—manganese, total dissolved solids, iron, sulfate, and chloride—were detected at high concentrations in 18.6, 8.6, 7.1, 2.9, and 1.4 percent of the primary aquifers, respectively. Of the 205 organic and special interest constituents analyzed, 32 constituents were detected. The herbicide simazine was the only organic constituent frequently detected, in 18.7 percent of grid wells, but at low relative-concentration.

The second component of this work, the *understanding assessment*, identified some of the primary natural and human factors that affect groundwater quality by evaluating correlations between land use, physical characteristics of the wells, geochemical conditions of the aquifer, water temperature, and relative-concentrations of constituents. Results from these analyses attempt to explain the occurrence and distribution of constituents in the MS study unit.

The *understanding assessment* indicated that wells that contained nitrate were significantly correlated with percentage agricultural land use, and had top perforations depths less than 350 feet (76 m). High and moderate relative-concentrations of arsenic may be attributed to reductive dissolution of manganese or iron oxides, and arsenic concentrations increased with groundwater age. Simazine was observed predominantly in wells with surrounding land use classified as agricultural or urban, and top of perforation depths less than 200 feet (61 m).

Acknowledgments

The authors thank the following cooperators for their support: the State Water Resources Control Board, LLNL, CDPH, and CDWR. We especially thank the cooperating well owners and water purveyors for their generosity in allowing the USGS to collect samples from their wells. Funding for this work was provided by State of California bonds authorized by Proposition 50 and administered by the State Water Board.

References

Aeschbach-Hertig, W., Peeters, F., Beyerle, U., and Kipfer, R., 1999, Interpretation of dissolved atmospheric noble gases in natural waters: Water Resources Research, v. 35, no. 9, p. 2779–2792.

Aeschbach-Hertig, W., Peeters, F., Beyerle, U., and Kipfer, R., 2000, Paleotemperature reconstruction from noble gases in groundwater taking into account equilibration with entrapped air: Nature, v. 405, p. 1040–1044.

Allen, J.E., 1946, Geology of the San Juan Bautista quadrangle: California Division of Mines Bulletin 133, 112 p.

Andrews, J.N., 1985, The isotopic composition of radiogenic helium and its use to study groundwater movement in confined aquifers: Chemical Geology, v. 49, p. 339–351.

Andrews, J.N., and Lee, D.J., 1979, Inert gases in groundwater from the Bunter Sandstone of England as indicators of age and paleoclimatic trends: Journal of Hydrology, v. 41, p. 233–252.

Ballantyne, J.M., and Moore, J.N., 1988, Arsenic geochemistry in geothermal systems: Geochimica et Cosmochimica Acta, v. 52, p. 475–483.

Belitz, Kenneth, Dubrovsky, N.M., Burow, K.R., Jurgens, B.C., and Johnson, T., 2003, Framework for a groundwater quality monitoring and assessment program for California: U.S. Geological Survey Water-Resources Investigations Report 03–4166, 28 p. (Also available at http://pubs.usgs. gov/wri/wri034166/.)

Belitz, K., Jurgens, B., Landon, M.K., Fram, M.S., and Johnson, T., 2010, Estimation of aquifer-scale proportion using equal-area grids: Assessment of regional-scale groundwater quality: Water Resources Research, v. 46, W11550, 14 p., doi:10.1029/2010WR009321. (Also available at http://www.agu.org/pubs/ crossref/2010/2010WR009321.shtml.)

Brown, L.D., Cai, T.T., and DasGupta, A., 2001, Interval estimation for a binomial proportion: Statistical Science, v. 16, no. 2, p. 101–117. (Also available at http://www.jstor. org/stable/2676784.)

California Department of Public Health, 2008a, Perchlorate in drinking water: California Department of Public Health website, accessed October 17, 2009, at http://www.cdph. ca.gov/certlic/drinkingwater/Pages/Perchlorate.aspx.

California Department of Public Health, 2008b, California drinking water—NDMA- related activities: California Department of Public Health website, accessed October 17, 2009, at http://www.cdph.ca.gov/certlic/drinkingwater/ Pages/NDMA.aspx.

California Department of Public Health, 2008c, 1,2,3-Trichloropropane: California Department of Public Health website, accessed October 17, 2009, at http://www. cdph.ca.gov/certlic/drinkingwater/Pages/123TCP.aspx.

California Department of Water Resources, 1977, North Monterey water resources investigation: Central district report to the Monterey County Flood Control and Water Conservation District, 20 p.

California Department of Water Resources, 1999, Evaluation of groundwater overdraft in the Southern Central Coast Region, Part 2. Technical Information Record SD-99-2, 116 p.

California Department of Water Resources, 2003, California's groundwater: California Department of Water Resources Bulletin 118, 246 p. (Also available at http://www.water. ca.gov/groundwater/bulletin118/update2003.cfm.)

California Environmental Protection Agency, 2010, GAMA— Groundwater Ambient Monitoring and Assessment Program: State Water Resources Control Board website, accessed September 9, 2010, at http://www.swrcb.ca.gov/ gama.

California State Water Resources Control Board, 2003, A comprehensive groundwater quality monitoring program for California: Assembly Bill 99 Report to the Governor and Legislature, March 2003, 100 p.

Chapelle, F.H., 2001, Groundwater microbiology and geochemistry (2nd ed.): New York, John Wiley and Sons, Inc., 477 p.

Chapelle, F.H., McMahon, P.B., Dubrovsky, N.M., Fuji, R.F., Oaksford, E.T., and Vroblesky, D.A., 1995, Deducing the distribution of terminal electron-accepting processes in hydrologically diverse groundwater systems: Water Resources Research, v. 31, no. 2, p. 359–371.

Clark, I.D., and Fritz, P., 1997, Environmental isotopes in hydrogeology: New York, Lewis Publishers, 328 p.

Cook, P.G., and Böhlke, J.K., 2000, Determining timescales for groundwater flow and solute transport, in Cook, P.G., and Herczeg, A., eds., Environmental tracers in subsurface hydrology: Boston, Kluwer Academic Publishers, p. 1–30.

Craig, H., and Lal, D., 1961, The production rate of natural tritium: Tellus, v. 13, p. 85–105.

Davis, S., and DeWiest, R.J., 1966, Hydrogeology: New York, John Wiley and Sons, 413 p.

Domagalski, J.L, and Dubrovsky, N.M., 1991, Regional assessment of non point-source pesticide residues in ground water, San Joaquin Valley, California: U.S. Geological Survey Water-Resources Investigations Report 91–4027, 14 p.

Dotsika, E., Poutoukis D., Michelot, J.L., and Kloppmann, W., 2006, Stable isotope and chloride, boron study for tracing sources of boron contamination in groundwater: Boron contents in fresh and thermal water in different areas in Greece: Water, Air, and Soil Pollution, v.174, p. 19–32.

Durbin, T.J., Kapple, G.W., and Freckleton, J.R., 1978, Two-dimensional and three-dimensional digital flow models of the Salinas Valley Groundwater Basin, California: U.S. Geological Survey Water-Resources Investigations Report 78–113, 134 p.

Durham, D.L, 1974, Geology of the Southern Salinas Valley Area, California, U.S. Geological Survey Professional Paper 819, Plate 3.

Fontes, J.C., and Garnier, J.M., 1979, Determination of the initial 14C activity of the total dissolved carbon: a review of the existing models and a new approach: Water Resources Research, v. 15, p. 399–413.

Frankenberger, W.T., ed., 2002, Environmental chemistry of arsenic: New York, Marcel Dekker, 391 p.

Gilliom, R.J., Barbash, J.E., Crawford, C.G., Hamilton, P.A., Martin, J.D., Nakagaki, N., Nowell, L.H., Scott, J.C., Stackelberg, P.E., Thelin, G.P., and Wolock, D.M., 2006, The quality of our nation's waters—Pesticides in the nation's streams and ground water, 1992–2001: U.S. Geological Survey Circular 1291, 172 p.

Greene, H.G., 1970, Geology of southern Monterey Bay and its relationship to the groundwater basin and seawater intrusion: U.S. Geological Survey Open-File Report 1465, 50 p.

Hanson, R.T., Everett, R.R., Newhouse, M.W., Crawford, S.M., Pimentel, M.I., and Smith, G.A., 2002, Geohydrology of a deep-aquifer system monitoring-well site at Marina, Monterey County, California: U.S. Geological Survey Water-Resources Investigation 02–4003, 73 p. (Also available at http://pubs.er.usgs.gov/publication/wri024003.)

Helsel, D.R., and Hirsch, R.M., 2002, Statistical methods in Water Resources: U.S. Geological Survey Techniques of Water-Resources Investigations, book 4, chap. A3, 510 p. (Also available at http://water.usgs.gov/pubs/twri/twri4a3/.)

Hem, J.D., 1970, Study and interpretation of the chemical characteristics of natural water (2d ed.): U.S. Geological Survey Water-Supply Paper 1473, 363 p.

Hutson, S.S., Barber, N.L., Kenny, J.F., Linsey, K.S., Lumia, D.S., and Maupin, M.A., 2004, Estimated use of water in the United States in 2000: U.S. Geological Survey Circular 1268, 46 p.

Isaaks, E.H., and Srivastava, R. M., 1989, Applied Geostatistics: New York, Oxford University Press, 511 p.

Islam, F.S., Gault, A.G., Boothman, C., Polya, D.A., Charnock, J.M., Chatterjee, D., and Lloyd, J.R., 2004, Role of metal-reducing bacteria in arsenic release from Bengal delta sediments: Nature, v. 430, p. 68–71.

Johnson, M.J., Londquist, C.J., Laudon, J., and Mitten, H.T., 1988, Geohydrology and mathematical simulation of the Pajaro Valley Aquifer system, Santa Cruz and Monterey Counties, California: U.S. Geological Survey Water-Resources Investigations Report 87–4281, 62 p.

Johnson, T.D., and Belitz, Kenneth, 2009, Assigning land use to supply wells for the statistical characterization of regional groundwater quality—Correlating urban land use and VOC occurrence: Journal of Hydrology, v. 370, p. 100–108.

Jurgens, B.C., McMahon, P.B., Chapelle, F.H., and Eberts, S.M., 2009, An Excel® workbook for identifying redox processes in ground water: U.S. Geological Survey Open-File Report 2009–1004, 8 p. (Also available at http://pubs.usgs.gov/of/2009/1004/.)

Kapple, G.W., Mitten, H.T., and Durbin, T.J., 1984, Analysis of the Carmel Valley alluvial ground-water basin, Monterey County, California: Monterey Peninsula Water-Management District Report 83–4580, 45 p.

Kulongoski, J.T., and Belitz, Kenneth, 2004, Ground-water ambient monitoring and assessment program: U.S. Geological Survey Fact Sheet 2004–3088, 2 p. (Also available at http://pubs.usgs.gov/fs/2004/3088/.)

Kulongoski, J.T., and Belitz, Kenneth, 2007, Ground-water quality data in the Monterey Bay and Salinas Valley Basins, California, 2005—Results from the California GAMA Program: U.S. Geological Survey Data Series 258, 84 p. (Also available at http://pubs.usgs.gov/ds/2007/258/.)

Kulongoski, J.T., Belitz, Kenneth, Landon, M.K., and Farrar, Christopher, 2010, Status and understanding of groundwater quality in the North San Francisco Bay Groundwater Basins, 2004: California GAMA Priority Basin Project: U.S. Geological Survey Scientific Investigations Report 2010–5089, 87 p. (Also available at http://pubs.usgs.gov/sir/2010/5089.)

Kulongoski, J.T., Hilton, D.R., Cresswell, R.G., Hostetler, S., and Jacobson, G., 2008, Helium-4 characteristics of groundwaters from Central Australia—Comparative chronology with chlorine-36 and carbon-14 dating techniques: Journal of Hydrology, v. 348, issues 1–2, p. 176–194, doi:10.1016/j.jhydrol.2007.09.048, accessed December 28, 2010, at http://www.sciencedirect.com/science/article/B6V6C-4PT0Y94-4/2/7e604b2f53860853a473b4213aa02028.

Landon, M.K., Belitz, Kenneth, Jurgens, B.C., Kulongoski, J.T., and Johnson, T.D., 2010, Status and understanding of groundwater quality in the Central–Eastside San Joaquin Basin, 2006—California GAMA Priority Basin project: U.S. Geological Survey Scientific Investigations Report 2009–5266, 97 p. (Also available at http://pubs.usgs.gov/sir/2009/5266/.)

Lindberg, R.D., and Runnells, D.D., 1984, Groundwater redox reactions: Science, v. 225, p. 925–927.

Lucas, L.L., and Unterweger, M.P., 2000, Comprehensive review and critical evaluation of the half-life of tritium: Journal of Research of the National Institute of Standards and Technology, v. 105, no. 4, p. 541–549.

Manning, A.H., Solomon, D.K., and Thiros, S.A., 2005, ^{3}H/^{3}He age data in assessing the susceptibility of wells to contamination: Ground Water, v. 43, no. 3, p. 353–367, doi:10.1111/j.1745-6584.2005.0028.x, accessed December 28, 2010, at http://onlinelibrary.wiley.com/doi/10.1111/j.1745-6584.2005.0028.x/full.

McMahon, P.B., and Chapelle, F.H., 2008, Redox processes and water quality of selected principal aquifer systems: Ground Water, v. 46, no. 2, p. 259–271. doi:10.1111/j.1745-6584.2007.00385.x, accessed December 28, 2010, at http://onlinelibrary.wiley.com/doi/10.1111/j.1745-6584.2007.00385.x/full.

Michel, R.L., 1989, Tritium deposition in the continental United States, 1953–83: U.S. Geological Survey Water-Resources Investigations Report 89–4072, 46 p.

Michel, R.L., and Schroeder, R., 1994, Use of long-term tritium records from the Colorado River to determine timescales for hydrologic processes associated with irrigation in the Imperial Valley, California: Applied Geochemistry, v. 9, p. 387–401.

Montgomery-Watson Consulting Engineers, 1994, Salinas River Basin water resources management plan task 1.09 Salinas Valley groundwater flow and quality model report: Prepared for Monterey County Water-Resources Agency, 33 p.

Morrison, P., and Pine, J., 1955, Radiogenic origin of the helium isotopes in rock: Annals of the New York Academy of Sciences, v. 12, p. 19–92.

Muir, K.S., 1980, Seawater intrusion and potential yield of aquifers in the Soquel-Aptos area, Santa Cruz County, California: U.S. Geological Survey Water Resources Investigations Report 80–84, 29 p.

Muir, K.S., 1982, Ground water in the Seaside area, Monterey County, California: U.S. Geological Survey Water Resources Investigations Report 82–10, 37 p.

Nakagaki, N., Price, C.V., Falcone, J.A., Hitt, K.J., and Ruddy, B.C., 2007, Enhanced National Land Cover Data 1992 (NLCDe 92): U.S. Geological Survey Raster digital data, accessed December 8, 2010, at http://water.usgs.gov/lookup/getspatial?nlcde92.

Nakagaki, N., and Wolock, D.M., 2005, Estimation of agricultural pesticide use in drainage basins using land cover maps and county pesticide data: U.S. Geological Survey Open-File Report 2005–1188, 56 p. (Also available at http://pubs.usgs.gov/of/2005/1188/.)

Piper, A.M., 1944, A graphic procedure in the geochemical interpretation of water analyses: American Geophysical Union Transactions, v. 25, p. 914–923.

Plummer, L.N., Michel, R.L., Thurman, E.M., and Glynn, P.D., 1993, Environmental tracers for age-dating young ground water, in Alley, W.M., ed., Regional Groundwater Quality: New York, Van Nostrand Reinhold, p. 255–294.

Poreda, R.J., Cerling, T.E., and Salomon, D.K., 1988, Tritium and helium isotopes as hydrologic tracers in a shallow unconfined aquifer: Journal of Hydrology, v. 103, p. 1–9.

Ravenscroft, P., Brammer, H., and Richards, K., 2009, Arsenic Pollution—A global synthesis: West Sussex, United Kingdom, Wiley-Blackwell, 588 p.

Reimann, C., and de Caritat, P., 1998, Chemical elements in the environment—Factsheets for the Geochemist and Environmental Scientist: Berlin, Springer-Verlag, 398 p.

Rowe, B.L., Toccalino, P.L., Moran, M.J., Zogorski, J.S., and Price, C.V., 2007, Occurrence and potential human-health relevance of volatile organic compounds in drinking water from domestic wells in the United States: Environmental Health Perspectives, v. 115, no. 11, p. 1539–1546, doi:10.1289/ehp.10253, accessed December 28, 2010, at http://ehp03.niehs.nih.gov/article/info%3Adoi%2F10.1289%2Fehp.10253.

Scott, J.C., 1990, Computerized stratified random site selection approaches for design of a ground-water-quality sampling network: U.S. Geological Survey Water-Resources Investigations Report 90–4101, 109 p.

Showalter, P., Akers, J.P., and Swain, L.A., 1983, Design of a ground-water quality monitoring network for the Salinas River Basin, California: U.S. Geological Survey Water-Resources Investigation Report 83–4049, 74 p.

Sparks, D.L., 1995, Environmental soil chemistry: San Diego, Academic Press, 353 p.

State of California, 1999, Supplemental Report of the 1999 Budget Act 1999–00 Fiscal Year, Item 3940-001-0001, State Water Resources Control Board, accessed September 9, 2010, at http://www.lao.ca.gov/1999/99-00_supp_rpt_lang_html#3940.

State of California, 2001a, Assembly Bill No. 599, Chapter 522, accessed September 9, 2010, at http://www.swrcb.ca.gov/gama/docs/ab_599_bill_20011005_chaptered.pdf.

State of California, 2001b, Groundwater Monitoring Act of 2001: California Water Code, part 2.76, Sections 10780–10782.3, accessed September 9, 2010, at http://www.leginfo.ca.gov/cgi-bin/displaycode?section=wat&group=10001-11000&file=10780-10782.3.

Stollenwerk, K., 2003, Geochemical processes controlling transport of arsenic, in Welch, A.H., and Stollenwerk, K.G., eds., Arsenic in ground water: Boston, Kluwer Academic Publishers, p. 475.

Takaoka, N., and Mizutani, Y., 1987, Tritiogenic ^{3}He in groundwater in Takaoka: Earth and Planetary Science Letters, v. 85, p. 74–78.

Toccalino, P.L., and Norman, J.E., 2006, Health-based screening levels to evaluate U.S. Geological Survey ground-water quality data: Risk Analysis, v. 26, no. 5, p. 1339–1348.

Toccalino, P.L., Norman, J.E., Phillips, R.H., Kauffman, L.J., Stackelberg, P.E., Nowell, L.H., Krietzman, S.J., and Post, G.B., 2004, Application of health-based screening levels to groundwater quality data in a state-scale pilot effort: U.S. Geological Survey Scientific Investigations Report 2004–5174, 64 p. (Also available at http://pubs.usgs.gov/sir/2004/5174/.)

Tolstikhin, I.N., and Kamenskiy, I.L., 1969, Determination of groundwater ages by the T-^{3}He method: Geochemistry International, v. 6, p. 310–811.

Torgersen, T., 1980, Controls on pore-fluid concentrations of ^{4}He and ^{222}Rn and the calculation of ^{4}He/^{222}Rn ages: Journal of Geochemical Exploration, v. 13, p. 7–75.

Torgersen, T., and Clarke, W.B., 1985, Helium accumulation in groundwater—I. An evaluation of sources and continental flux of crustal ^{4}He in the Great Artesian basin, Australia: Geochimica et Cosmochimica Acta, v. 49, p. 1211–1218.

Torgersen, T., Clarke, W.B., and Jenkins, W.J., 1979, The tritium/helium-3 method in hydrology, in Isotope Hydrology 1978: IAEA-SM-228/2, IAEA, Vienna, p. 917–930.

Troiano, J., Weaver, D., Marade, J., Spurlock, F., Pepple, M., Nordmark, C., and Bartkowiak, D., 2001, Summary of well water sampling in California to detect pesticide residues resulting from nonpoint source applications: Journal of Environmental Quality, v. 30, p. 448–459.

U.S. Environmental Protection Agency, 1998, Reporting requirements for risk/benefit information: Federal Register, v. 62, no. 182, p. 49369–49395, accessed October 17, 2009, at http://www.epa.gov/EPA-PEST/1997/September/Day-19/p24937.htm.

U.S. Environmental Protection Agency, 2006, 2006 Edition of the drinking water standards and health advisories: Washington, D.C., U.S. Environmental Protection Agency, Office of Water EPA/822/R-06-013 (Also available at http://www.epa.gov/waterscience/criteria/drinking/dwstandards.pdf.)

U.S. Geological Survey, 2010, What is the Priority Basin Project?: U.S. Geological Survey website, accessed September 9, 2010, at http://ca.water.usgs.gov/gama.

Vogel, J.C., and Ehhalt, D., 1963, The use of the carbon isotopes in groundwater studies—Radioisotopes in Hydrology: Vienna, IAEA, p. 383–395.

Webster, J.G., and Nordstrom, D.K., 2003, Geothermal arsenic, in Welch, A.H., and Stollenwerk, K.G., eds., Arsenic in ground water: Boston, Kluwer Academic Publishers, p. 101–126.

Welch, A.H., Lico, M.S., and Hughes, J.L., 1988, Arsenic in ground water of the western United States: Ground Water, v. 26, no. 3, p. 333–347.

Welch, A.H., Oremland, R.S., Davis, J.A., and Watkins, S.A., 2006, Arsenic in ground water—A review of current knowledge and relation to the CALFED solution area with recommendations for needed research: San Francisco Estuary and Watershed Science, v. 4, no. 2, Article 4, 32 p., accessed May 19, 2008, at http://repositories.cdlib.org/jmie/sfews/vol4/iss2/art4/.

Welch, A.H., Westjohn, D.B., Helsel, D.R., and Wanty, R.B., 2000, Arsenic in ground water of the United States—occurrence and geochemistry: Ground Water, v. 38, no. 4, p. 589–604.

Zogorski, J.S., Carter, J.M., Ivahnenko, Tamara, Lapham, W.W., Moran, M.J., Rowe, B.L., Squillace, P.J., and Toccalino, P.L., 2006, The quality of our Nation's waters—Volatile organic compounds in the Nation's ground water and drinking-water supply wells: U.S. Geological Survey Circular 1292, 101 p.

Appendix A. Ancillary Datasets

Land-Use Classification

Land use was classified by using an enhanced version of the satellite derived (98 ft (30 m) pixel resolution) USGS National Land Cover Dataset (Nakagaki and others, 2007). This dataset has been used in previous national and regional studies relating land use to water quality (Gilliom and others, 2006; Zogorski and others, 2006). The dataset characterizes land cover during the early 1990s. One pixel in the dataset imagery represents a land area of 9,688 ft² (900 m²), calculated from the pixel of 98 ft (30 m). The imagery was classified into 25 land-cover classifications (Nakagaki and Wolock, 2005). These 25 land-cover classifications were aggregated into four principal land-use classes—urban, agricultural, natural, and mixed. Each pixel was assigned a land-use class if greater than 50 percent of the land cover in that area could be associated with a single land use. If no land cover was greater than 50 percent of the pixel area, the classification of "mixed" was assigned.

Land-use classes for the study unit, for study areas, and for areas within a radius of 1,640 ft (500 m) surrounding each well were assigned using the USGS National Land Cover Dataset (Johnson and Belitz, 2009). Land-use classes for the study unit and the study areas (fig. 6) were calculated from the land cover of each pixel in the study unit and the study areas. Land use assigned to the area surrounding an individual well (table A1) was calculated from land use within the area surrounding each well (radius of 1,640 ft (500 m) and land area of 8,449,620 ft² (785,400 m²)). For some analyses of constituent distributions, urban and agricultural land-use classes were combined into a single class urban/agriculture to represent land used for anthropogenic purposes (fig. 15B).

Well Construction Information

Most well-construction data were from driller's logs and are given in table A1. Other sources were ancillary records of well owners and the USGS National Water Information System database. Well identification verification procedures are described by Kulongoski and Belitz (2007).

Normalized Position of Wells Along a Flowpath

The normalized position of wells in the Salinas River Valley in relation to the groundwater flow system was an additional factor examined for the understanding of water quality in the MS study unit (table A1). The flowpath considered in this study extended from the upper Salinas Valley to Monterey Bay. Groundwater in the alluvium moves under a natural hydraulic gradient that conforms in a general way to the surface topography. In the Salinas Valley, groundwater movement generally is from the southern part of the valley northward towards Monterey Bay (fig. 8). Normalized position along the flowpath was determined by calculating the valley length (the distance from the southernmost point in the valleys to Monterey Bay). Then a perpendicular line was drawn from each well to the valley upgradient-downgradient axis (typically the location of the river), demarking the normalized position, or distance along the flowpath. Positions were normalized by dividing the distance of the projected location along the flowpath by the total length of the system, resulting in a value from 0 to 1; normalized positions are given in table A1. Low values of normalized position indicate locations in the upgradient or proximal portion of the flow system and high values of position indicate locations in the downgradient or distal portion of the flow system. Plotting data with respect to normalized position along the flowpath also allows for aggregation of areally distributed data into a single diagrammatic cross section.

Classification of Geochemical Condition

Geochemical conditions investigated as potential explanatory variables in this report include oxidation-reduction characteristics, dissolved oxygen, and ratios of iron, arsenic, and chromium species (table A2). An automated workbook program was used to assign the redox classification to each sample (Jurgens and others, 2009). Oxidation-reduction (redox) conditions influence the mobility of many organic and inorganic constituents (McMahon and Chapelle, 2008). Along groundwater flowpaths, redox conditions commonly proceed along a well-documented sequence of Terminal Electron Acceptor Processes (TEAP); one TEAP typically dominates at a particular time and aquifer location (Chapelle and others, 1995; Chapelle, 2001). The predominant TEAPs are oxygen-reduction, nitrate-reduction, manganese-reduction, iron-reduction, sulfate-reduction, and methanogenesis. The presence of redox-sensitive chemical species suggesting more than one TEAP may indicate mixed waters from different redox zones upgradient of the well, a well screened across more than one redox zone, or spatial heterogeneity in microbial activity in the aquifer. Different redox elements (for example; iron, manganese, and sulfur) tend not to reach overall equilibrium in most natural water systems (Lindberg and Runnels, 1984); therefore, a single redox measurement usually cannot represent the system, further complicating the assessment of redox conditions. pH is the measure of hydrogen-ion activity in a water sample and is sensitive to redox conditions.

Groundwater-Age Classification

Groundwater recharge temperature from noble gases, age data, and classifications are listed in table A3. Groundwater dating techniques indicate the time since the groundwater was last in contact with the atmosphere. Techniques used to estimate groundwater residence times or 'age' include those based on tritium (for example, Tolstikhin and Kamenskiy, 1969; Torgersen and others, 1979), tritium combined with its decay product helium-3 (for example, Takaoka and Mizutani, 1987; Poreda and others, 1988), carbon-14 activities (for example, Vogel and Ehhalt, 1963; Plummer and others, 1993), and dissolved noble gases, particularly helium-4 accumulation (for example, Davis and DeWiest, 1966; Andrews and Lee, 1979; Kulongoski and others, 2008).

Tritium (^{3}H) is a short-lived radioactive isotope of hydrogen with a half-life of 12.32 years (Lucas and Unterweger, 2000). Tritium is produced naturally in the atmosphere from the interaction of cosmogenic radiation with nitrogen (Craig and Lal, 1961), by above-ground nuclear explosions, and by the operation of nuclear reactors. Tritium enters the hydrological cycle following oxidation to tritiated water (HTO). Consequently, the presence of ^{3}H in groundwater may be used to identify water that has exchanged with the atmosphere since 1953. By determining the ratio of ^{3}H to ^{3}He, resulting from the radioactive decay of ^{3}H, the time that the water has resided in the aquifer can be calculated more precisely than by using tritium alone (for example, Takaoka and Mizutani, 1987; Poreda and others, 1988).

Carbon-14 (^{14}C) is a widely used chronometer based on the radiocarbon content of dissolved inorganic carbonate species in groundwater. ^{14}C is formed in the atmosphere by the interaction of cosmic-ray neutrons with nitrogen and, to a lesser degree, with oxygen and carbon. ^{14}C is incorporated into carbon dioxide and mixed throughout the atmosphere, dissolved in precipitation, and incorporated into the hydrologic cycle. ^{14}C activity in groundwater, expressed as percent modern carbon (pmc), reflects exposure to the atmospheric ^{14}C source and is governed by the decay constant of ^{14}C (with a half-life of 5,730 years). ^{14}C can be used to estimate groundwater ages ranging from 1,000 to less than 30,000 years before present because of its half-life. Calculated ^{14}C ages in this study are referred to as "uncorrected" because they have not been adjusted to consider exchanges with sedimentary sources of carbon (Fontes and Garnier, 1979). The ^{14}C age (residence time) is calculated on the basis of the decrease in ^{14}C activity as a result of radioactive decay since groundwater recharge, relative to an assumed initial ^{14}C concentration (Clarke and Fritz, 1997). A mean initial ^{14}C activity of 99 percent modern carbon (pmc) was assumed for this study, with estimated errors on calculated groundwater ages of as much as ±20 percent.

Helium (He) is a naturally occurring inert gas initially included during the accretion of the planet, and later produced by the radioactive decay of lithium, thorium, and uranium in the Earth. Measured He concentrations in groundwater is the sum of several He components including air-equilibrated He (He_{eq}), He from dissolved-air bubbles (He_a), terrigenic He (He_{ter}), and tritiogenic He-3 (^3He_t). Helium (^{3}He and ^{4}He) concentrations in groundwater often exceed the expected solubility equilibrium values, a function of the temperature of the water, as a result of subsurface production of both isotopes and their subsequent release into the groundwater (for example, Morrison and Pine, 1955; Andrews and Lee, 1979; Torgersen, 1980; Andrews, 1985; Torgersen and Clarke, 1985). The presence of terrigenic He in groundwater, from its production in aquifer material or deeper in the crust, is indicative of long groundwater residence times. The amount of terrigenic helium is defined as the concentration of the total measured helium minus the fraction as a result of air-equilibration [He_{eq}] and dissolved air-bubbles [He_a]. For the purposes of this study, percent terrigenic He is used to identify groundwater with residence times greater than 100 years. Percent terrigenic He is defined as the concentration of terrigenic He (as defined previously) divided by the total measured He in the sample (corrected for air-bubble entrainment). Samples with greater than 5 percent terrigenic He indicate that groundwater has a residence time of more than 100 years.

Recharge temperatures for 96 samples were calculated from dissolved neon, argon, krypton, and xenon data by using methods described by Aeschbach-Hertig and others (1999). The only modeled recharge temperatures accepted were those for which the probability was greater than 1 percent that the sum of the squared deviations between the modeled and the measured concentrations (weighted with the experimental 1-sigma errors) was equal to or greater than the observed value (Aeschbach-Hertig and others, 2000). The recharge temperature with the highest probability for each sample was used in this report.

^{3}H/^{3}He ages were computed as described by Poreda and others (1988). The ^{3}He/^{4}He of samples was determined by the linear regression of the percentage of terrigenic He and δ^3He [(δ^3He = R_{meas}/R_{atm} −1) × 100 percent] of samples containing less than 1 tritium unit. Calculations of the noble gas temperature and ^{3}He to ^{4}He ratios are useful because they constrain helium-based groundwater ages further.

In this study, the age distributions of samples are classified as pre-modern, modern, and mixed. Groundwater with tritium activity less than 1 tritium unit (TU), percent terrigenic He greater than 5 percent, and ^{14}C less than 90 pmc was designated as pre-modern, defined as having been recharged before 1953. Groundwater with tritium activities greater than 1 TU, percent terrigenic He less than 5 percent, and ^{14}C greater than 90 pmc is designated as modern, defined as having been recharged after 1953. Samples with pre-modern and modern components are designated as mixed groundwater, which includes substantial fractions of old and young waters. In reality, pre-modern groundwater could contain small fractions of modern water and modern water could contain small fractions of pre-modern water. Previous investigations have used a range of tritium values from 0.3 to 1.0 TU as thresholds for distinguishing pre-1953 from post-1953 water (Michel, 1989; Plummer and others, 1993; Michel and Schroeder, 1994; Clark and Fritz, 1997; Manning and others, 2005). By using a tritium value of 1.0 TU for the threshold in this study, the age classification scheme allows a slightly larger fraction of modern water to be classified as pre-modern than if a lower threshold were used. A lower threshold for tritium would result in fewer samples classified as pre-modern than as mixed, when other tracers, such as carbon-14 and terrigenic helium, would suggest that they were primarily pre-modern. This higher threshold was considered

more appropriate for this study because many of the wells were production wells with long screens and mixing of waters of different ages is likely to occur.

Tritium, percent modern carbon, and percent terrigenic helium, and sample-age classifications are reported in table A3. Because of uncertainties in age distributions, in particular those caused by mixing waters of different ages in wells with long perforation intervals and high withdrawal rates, these age estimates were not specifically used for statistically quantifying the relation between age and water quality in this report. Although more sophisticated lumped parameter models used for analyzing age distributions that incorporate mixing are available (for example, Cook and Böhlke, 2000), use of these alternative models to characterize age mixtures was beyond the scope of this report. Rather, classification into modern (recharged after 1953), mixed, and pre-modern (recharged before 1953) categories was sufficient to provide an appropriate and useful characterization for the purposes of examining groundwater quality.

Table A1. Percent land use by category, land use classification, well construction information, the normalized position of selected wells along a flowpath, cell number and USGS GAMA well identification number for well data used in the Monterey Bay and Salinas Valley Basins study unit, California GAMA Priority Basin Project.

[USGS-GAMA well identification number: MB, Monterey Bay study area well;PR, Paso Robles study area well; SC, Santa Cruz study area well; SV, Salinas Valley study area well. **Data source:** DG, use of CDPH inorganic constituent data at a USGS-grid well; DPH, a CDPH-grid well with no USGS data; FP, flow-path well; no CDPH data, no CDPH water-quality data available; no USGS well, USGS GAMA-grid well was not sampled in that cell; USGS data, USGS-GAMA data available for the grid well. **Well type:** DOM, domestic; IRR, irrigation; MW, monitoring; PSW, public supply. **Construction information:** Well depth and top and bottom of perforations measured in feet below land surface datum (LSD). Length is from the top of the uppermost perforated interval to the bottom of the well. **Normalized position of well along flowpath** is dimensionless; measured on a scale of 0 to 1, with 0 being the most proximal (upgradient) position and 1 being the most distal (downgradient) position along the flowpath. nd, no data; na, not available or not applicable]

Grid cell No.	USGS-GAMA well identification number	Data source	Well type	Land use within 500 meters of the well, in percent			Land-use classification	Construction information, in feet				Normalized position of well along flowpath
				Agricultural	Natural	Urban		Well depth	Depth to top of perforations	Depth to bottom of perforations	Length	
Monterey Bay study area grid wells												
1	MB-01	MB-DG-01	PSW	0.0	2.6	97.4	Urban	280	nd	nd	nd	na
1		MB-DPH-60	PSW	0.2	88.0	11.8	Natural	nd	nd	nd	nd	na
2	MB-02	MB-DG-02	PSW	0.0	9.0	91.0	Urban	228	188	218	30	na
2		MB-DPH-61	PSW	0.0	11.6	88.4	Urban	nd	nd	nd	nd	na
3	MB-03	No CDPH data	PSW	37.0	61.4	1.6	Natural	1,364	1,301	1,361	60	na
3		MB-DPH-50	PSW	0.0	83.7	16.3	Natural	nd	nd	nd	nd	na
4	MB-04	USGS data	PSW	50.1	17.8	32.2	Agricultural	800	200	800	600	na
5	MB-05	MB-DG-05	PSW	15.9	9.9	74.2	Urban	440	200	430	230	na
6	MB-06	MB-DG-06	PSW	0.0	18.2	81.8	Urban	600	240	585	345	na
7	MB-07	MB-DG-07	PSW	0.0	2.2	97.8	Urban	680	470	670	200	na
8	MB-08	No CDPH data	PSW	0.0	61.9	38.1	Natural	550	440	520	80	na
8		MB-DPH-51	PSW	0.0	6.3	93.7	Urban	nd	nd	nd	nd	na
9	MB-09	USGS data	PSW	2.6	4.1	93.2	Urban	466	198	446	248	na
10	MB-10	MB-DG-10	PSW	77.0	3.9	19.1	Agricultural	600	500	600	100	na
11	MB-11	MB-DG-11	PSW	38.3	4.9	56.8	Urban	600	228	600	372	na
12	MB-12	USGS data	PSW	0.0	78.7	21.3	Natural	1,950	1,390	1,940	550	1.000
13	MB-13	No CDPH data	PSW	0.0	91.8	8.2	Natural	557	315	535	220	0.910
13		MB-DPH-52	PSW	0.0	12.0	88.0	Urban	nd	nd	nd	nd	na
14	MB-14	MB-DG-14	PSW	0.2	80.9	18.9	Natural	460	160	440	280	na
15	MB-15	MB-DG-15	PSW	0.3	15.8	83.8	Urban	552	315	535	220	na
16	MB-16	No CDPH data	PSW	0.0	94.0	6.0	Natural	552	315	535	220	0.902
17	MB-17	MB-DG-17	PSW	33.3	15.1	51.5	Urban	630	370	610	240	na
18	MB-18	USGS data	PSW	13.7	43.8	42.5	Mixed	640	200	620	420	na
19	MB-19	MB-DG-19	PSW	14.9	45.2	39.9	Mixed	470	290	460	170	na
20	MB-20	USGS data	PSW	1.5	9.3	89.2	Urban	177	103	147	44	na
21	MB-21	No CDPH data	DOM	24.3	41.4	34.4	Mixed	200	nd	nd	nd	na

Table A1. Percent land use by category, land use classification, well construction information, the normalized position of selected wells along a flowpath, cell number and USGS GAMA well identification number for well data used in the Monterey Bay and Salinas Valley Basins study unit, California GAMA Priority Basin Project.—Continued

[USGS-GAMA well identification number: MB, Monterey Bay study area well; PR, Paso Robles study area well; SC, Santa Cruz study area well; SV, Salinas Valley study area well. Data source: DG, use of CDPH inorganic constituent data at a USGS-grid well; DPH, a CDPH-grid well with no USGS data; FP, flow-path well; no CDPH data, no CDPH water-quality data available; no USGS well, USGS GAMA-grid well was not sampled in that cell; USGS data, USGS-GAMA data available for the grid well. Well type: DOM, domestic; IRR, irrigation; MW, monitoring; PSW, public supply. Construction information: Well depth and top and bottom of perforations measured in feet below land surface datum (LSD). Length is from the top of the uppermost perforated interval to the bottom of the well. Normalized position of well along flowpath is dimensionless; measured on a scale of 0 to 1, with 0 being the most proximal (upgradient) position and 1 being the most distal (downgradient) position along the flowpath. nd, no data; na, not available or not applicable]

Grid cell No.	USGS-GAMA well identification number	Data source	Well type	Land use within 500 meters of the well, in percent			Land-use classification	Construction information, in feet				Normalized position of well along flowpath
				Agricultural	Natural	Urban		Well depth	Depth to top of perforations	Depth to bottom of perforations	Length	
Monterey Bay study area grid wells—Continued												
22	MB-22	USGS data	IRR	96.7	3.1	0.2	Agricultural	420	300	400	100	na
23	MB-23	MB-DG-23	PSW	0.8	22.3	76.9	Urban	280	240	280	40	na
23		MB-DPH-72	PSW	13.6	44.4	42.0	Mixed	nd	nd	nd	nd	na
24	MB-24	MB-DG-24	PSW	17.9	13.1	69.1	Urban	600	300	600	300	na
25	MB-25	MB-DG-25	PSW	82.1	6.4	11.5	Agricultural	nd	nd	nd	nd	na
26	MB-26	MB-DG-26	PSW	1.8	1.5	96.7	Urban	650	451	624	173	0.861
27	MB-27	MB-DG-27	PSW	0.0	43.6	56.4	Urban	610	290	610	320	na
27		MB-DPH-66	PSW	0.0	38.8	61.2	Urban	nd	nd	nd	nd	na
28	MB-28	MB-DG-28	PSW	54.4	41.4	4.2	Agricultural	490	413	465	52	na
28		MB-DPH-67	PSW	63.9	30.2	5.8	Agricultural	nd	nd	nd	nd	na
29	MB-29	USGS data	PSW	93.0	0.1	6.9	Agricultural	274	129	256	127	0.828
30	MB-30	USGS data	PSW	0.0	2.6	97.4	Urban	668	475	652	177	na
31	MB-31	MB-DG-31	PSW	47.8	3.7	48.6	Mixed	619	198	607	409	na
32	MB-32	MB-DG-32	PSW	0.0	17.8	82.2	Urban	214	nd	nd	nd	na
32		MB-DPH-68	PSW	19.7	31.2	49.1	Mixed	nd	nd	nd	nd	na
33	MB-33	USGS data	PSW	0.0	61.3	38.7	Natural	500	nd	nd	nd	na
34	MB-34	MB-DG-34	PSW	12.1	59.1	28.8	Natural	510	220	490	270	na
34		MB-DPH-69	PSW	25.3	30.6	44.1	Mixed	nd	nd	nd	nd	na
35	MB-35	USGS data	PSW	92.4	6.1	1.5	Agricultural	nd	nd	nd	nd	na
36	MB-36	MB-DG-36	PSW	93.1	3.2	3.7	Agricultural	640	392	620	228	na
37	MB-37	USGS data	PSW	26.1	14.3	59.6	Urban	810	310	810	500	na
38	MB-38	MB-DG-38	PSW	10.0	13.9	76.2	Urban	630	360	610	250	na
39	MB-39	No CDPH data	IRR	89.3	1.9	8.7	Agricultural	nd	nd	nd	nd	0.770
39		MB-DPH-54	PSW	86.0	5.7	8.2	Agricultural	nd	nd	nd	nd	na
40	MB-40	USGS data	PSW	79.3	7.9	12.8	Agricultural	nd	nd	nd	nd	0.790
41	MB-41	No CDPH data	IRR	99.5	0.5	0.0	Agricultural	nd	nd	nd	nd	0.740
41		MB-DPH-55	PSW	9.5	89.9	0.6	Natural	nd	nd	nd	nd	na

Table A1. Percent land use by category, land use classification, well construction information, the normalized position of selected wells along a flowpath, cell number and USGS GAMA well identification number for well data used in the Monterey Bay and Salinas Valley Basins study unit, California GAMA Priority Basin Project.—Continued

[USGS-GAMA well identification number: MB, Monterey Bay study area well; PR, Paso Robles study area well; SC, Santa Cruz study area well; SV, Salinas Valley study area well. Data source: DG, use of CDPH inorganic constituent data at a USGS-grid well; DPH, a CDPH-grid well with no USGS data; FP, flow-path well; no CDPH data, no CDPH water-quality data available; no USGS well, USGS GAMA-grid well was not sampled in that cell; USGS data, USGS-GAMA data available for the grid well. Well type: DOM, domestic; IRR, irrigation; MW, monitoring; PSW, public supply. Construction information: Well depth and top and bottom of perforations measured in feet below land surface datum (LSD). Length is from the top of the uppermost perforated interval to the bottom of the well. Normalized position of well along flowpath is dimensionless; measured on a scale of 0 to 1, with 0 being the most proximal (upgradient) position and 1 being the most distal (downgradient) position along the flowpath. nd, no data; na, not available or not applicable]

Grid cell No.	USGS-GAMA well identification number	Data source	Well type	Land use within 500 meters of the well, in percent			Land-use classification	Construction information, in feet				Normalized position of well along flowpath
				Agricultural	Natural	Urban		Well depth	Depth to top of perforations	Depth to bottom of perforations	Length	
Monterey Bay study area grid wells—Continued												
42	MB-42	No CDPH data	PSW	16.4	25.7	58.0	Urban	105	55	95	40	na
42		MB-DPH-56	PSW	59.2	9.4	31.4	Agricultural	nd	nd	nd	nd	na
43		MB-DG-43	PSW	48.6	10.7	40.8	Mixed	nd	nd	nd	nd	0.639
43		MB-DPH-71	PSW	42.4	16.6	41.0	Mixed	nd	nd	nd	nd	na
44	MB-44	USGS data	PSW	67.4	17.8	14.9	Agricultural	392	332	390	58	0.664
45	MB-45	USGS data	PSW	38.1	19.1	42.7	Mixed	148	55	128	73	na
46	MB-46	No CDPH data	IRR	93.6	6.4	0.0	Agricultural	610	350	610	260	0.705
46		MB-DPH-57	PSW	86.5	11.4	2.1	Agricultural	nd	nd	nd	nd	na
47	MB-47	USGS data	PSW	57.7	16.0	26.2	Agricultural	69	nd	nd	nd	0.721
48		MB-DG-48	PSW	80.5	7.6	11.9	Agricultural	nd	nd	nd	nd	na
Paso Robles study area grid wells												
1	PR-07	PR-DG-07	PSW	16.4	83.6	0.0	Natural	510	310	510	200	na
2	PR-09	PR-DG-09	PSW	0.2	96.1	3.7	Natural	502	72	495	423	na
3	PR-10	USGS data	PSW	0.1	88.7	11.2	Natural	201	nd	nd	nd	na
4	No USGS well	No CDPH data	nd	nd	nd	nd	nd	nd	nd	nd	nd	na
5	PR-08	USGS data	PSW	5.3	29.4	65.3	Urban	300	87	270	183	na
6	PR-06	PR-DG-06	PSW	36.9	31.8	31.3	Mixed	166	126	166	40	na
6		PR-DPH-30	PSW	35.1	52.0	12.9	Natural	nd	nd	nd	nd	na
7	PR-04	No CDPH data	PSW	64.3	30.9	4.8	Agricultural	775	275	775	500	na
7		PR-DPH-21	PSW	50.5	21.3	28.2	Agricultural	nd	nd	nd	nd	na
8	PR-02	PR-DG-02	PSW	74.8	12.9	12.3	Agricultural	290	120	290	170	na
9	PR-01	USGS data	PSW	60.0	37.1	2.9	Agricultural	500	150	500	350	na
10	PR-03	PR-DG-03	PSW	2.6	97.1	0.2	Natural	680	260	660	400	na
11	PR-05	PR-DG-05	PSW	26.0	69.3	4.7	Natural	840	640	840	200	na
12	No USGS well	PR-DPH-22	PSW	55.0	44.9	0.1	Agricultural	nd	nd	nd	nd	na
13	PR-11	No CDPH data	PSW	13.5	63.1	23.4	Natural	440	300	440	140	na
13		PR-DPH-23	PSW	13.5	63.1	23.4	Natural	nd	nd	nd	nd	na

Table A1. Percent land use by category, land use classification, well construction information, the normalized position of selected wells along a flowpath, cell number and USGS GAMA well identification number for well data used in the Monterey Bay and Salinas Valley Basins study unit, California GANA Priority Basin Project.—Continued

[USGS-GAMA well identification number:: MB, Monterey Bay study area well;PR, Paso Robles study area well; SC, Santa Cruz study area well; SV, Salinas Valley study area well. **Data source:** DG, use of CDPH inorganic constituent data at a USGS-grid well; DPH, a CDPH-grid well with no USGS data; FP, flow-path well; no CDPH data, no CDPH water-quality data available; no USGS well, USGS GAMA-grid well was not sampled in that cell; USGS data, USGS-GAMA data available for the grid well. **Well type:** DOM, domestic; IRR, irrigation; MW, monitoring; PSW, public supply. **Construction information:** Well depth and top and bottom of perforations measured in feet below land surface datum (LSD). Length is from the top of the uppermost perforated interval to the bottom of the well. **Normalized position of well along flowpath** is dimensionless; measured on a scale of 0 to 1, with 0 being the most proximal (upgradient) position and 1 being the most distal (downgradient) position along the flowpath. nd, no data; na, not available or not applicable]

Grid cell No.	USGS-GAMA well identification number	Data source	Well type	Land use within 500 meters of the well, in percent			Land-use classification	Construction information, in feet				Normalized position of well along flowpath
				Agricultural	Natural	Urban		Well depth	Depth to top of perforations	Depth to bottom of perforations	Length	
Paso Robles study area grid wells—Continued												
14	No USGS well	No CDPH data	nd	nd	nd	nd	nd	nd	nd	nd	nd	na
15	No USGS well	No CDPH data	nd	nd	nd	nd	nd	nd	nd	nd	nd	na
16	No USGS well	No CDPH data	nd	nd	nd	nd	nd	nd	nd	nd	nd	na
Salinas Valley study area grid wells												
1	SV-16	No CDPH data	PSW	67.6	21.9	10.5	Agricultural	800	200	790	590	0.525
2	SV-17	No CDPH data	PSW	87.6	11.5	0.9	Agricultural	nd	nd	nd	nd	0.550
3	No USGS well	No CDPH data	nd	nd	nd	nd	nd	nd	nd	nd	nd	na
4	SV-DPH-40	No CDPH data	PSW	74.6	12.4	13.1	Agricultural	nd	nd	nd	nd	na
4	SV-DPH-46	No CDPH data	PSW	77.8	11.3	10.9	Agricultural	nd	nd	nd	nd	na
5	SV-19	USGS data	PSW	70.6	13.3	16.2	Agricultural	830	500	816	316	0.590
6	No USGS well	No CDPH data	nd	nd	nd	nd	nd	nd	nd	nd	nd	na
7	SV-18	USGS data	PSW	69.9	7.1	23.0	Agricultural	nd	nd	nd	nd	0.525
8	No USGS well	No CDPH data	nd	nd	nd	nd	nd	nd	nd	nd	nd	na
9	SV-13	No CDPH data	IRR	86.0	14.0	0.0	Agricultural	210	140	200	60	0.459
10	SV-DPH-42	No CDPH data	PSW	90.1	9.9	0.0	Agricultural	nd	nd	nd	nd	na
11	No USGS well	No CDPH data	nd	nd	nd	nd	nd	nd	nd	nd	nd	na
12	SV-10	USGS data	PSW	89.8	7.2	3.0	Agricultural	nd	nd	nd	nd	0.377
13	SV-11	USGS data	PSW	68.8	4.8	26.3	Agricultural	883	313	863	550	0.451
14	SV-12	No CDPH data	IRR	88.4	11.6	0.0	Agricultural	nd	nd	nd	nd	0.459
14	SV-DPH-43	No CDPH data	PSW	51.1	7.3	41.6	Agricultural	nd	nd	nd	nd	na
15	SV-14	No CDPH data	IRR	93.6	6.4	0.0	Agricultural	235	106	232	126	0.467
16	SV-15	No CDPH data	IRR	58.9	41.1	0.0	Agricultural	200	60	180	120	0.471
16	SV-44	No CDPH data	PSW	54.4	45.5	0.1	Agricultural	nd	nd	nd	nd	na
17	SV-09	No CDPH data	IRR	79.8	16.8	3.3	Agricultural	275	115	220	105	0.369
17	SV-45	No CDPH data	PSW	80.0	15.2	4.8	Agricultural	nd	nd	nd	nd	na
18	SV-08	No CDPH data	IRR	94.7	5.3	0.0	Agricultural	nd	nd	nd	nd	0.328

Table A1. Percent land use by category, land use classification, well construction information, the normalized position of selected wells along a flowpath, cell number and USGS GAMA well identification number for well data used in the Monterey Bay and Salinas Valley Basins study unit, California GAMA Priority Basin Project.—Continued

[USGS-GAMA well identification number: MB, Monterey Bay study area well; PR, Paso Robles study area well; SC, Santa Cruz study area well; SV, Salinas Valley study area well. Data source: DG, use of CDPH inorganic constituent data at a USGS-grid well; DPH, a CDPH-grid well with no USGS data; FP, flow-path well; no CDPH data, no CDPH water-quality data available; no USGS well, USGS GAMA-grid well was not sampled in that cell; USGS data, USGS GAMA data available for the grid well. Well type: DOM, domestic; IRR, irrigation; MW, monitoring; PSW, public supply. Construction information: Well depth and top and bottom of perforations measured in feet below land surface datum (LSD). Length is from the top of the uppermost perforated interval to the bottom of the well. Normalized position of well along flowpath is dimensionless; measured on a scale of 0 to 1, with 0 being the most proximal (upgradient) position and 1 being the most distal (downgradient) position along the flowpath. nd, no data; na, not available or not applicable]

Grid cell No.	USGS-GAMA well identification number	Data source	Well type	Land use within 500 meters of the well, in percent			Land-use classifi-cation	Construction information, in feet				Normalized position of well along flowpath
				Agricultural	Natural	Urban		Well depth	Depth to top of perforations	Depth to bottom of perforations	Length	
Salinas Valley study area grid wells—Continued												
19	SV-07	USGS data	PSW	0.0	40.9	59.1	Urban	212	130	202	72	0.295
20	SV-06	SV-DG-06	PSW	53.7	45.1	1.1	Agricultural	220	160	220	60	0.287
21	SV-04	No CDPH data	DOM	97.6	2.4	0.0	Agricultural	432	120	424	304	0.246
22	No USGS well	No CDPH data	nd	nd	nd	nd	nd	nd	nd	nd	nd	na
23	SV-03	USGS data	PSW	85.1	14.4	0.5	Agricultural	140	90	140	50	0.180
24	No USGS well	No CDPH data	nd	nd	nd	nd	nd	nd	nd	nd	nd	na
25	SV-05	SV-DG-05	PSW	62.1	11.8	26.1	Agricultural	190	140	180	40	0.262
26	No USGS well	No CDPH data	nd	nd	nd	nd	nd	nd	nd	nd	nd	na
27	No USGS well	No CDPH data	nd	nd	nd	nd	nd	nd	nd	nd	nd	na
28	No USGS well	No CDPH data	nd	nd	nd	nd	nd	nd	nd	nd	nd	na
29	SV-02	USGS data	PSW	62.4	33.2	4.4	Agricultural	130	80	130	50	0.074
30	No USGS well	No CDPH data	nd	nd	nd	nd	nd	nd	nd	nd	nd	na
31	SV-01	USGS data	PSW	0.0	61.5	38.5	Natural	nd	nd	nd	na	0.008
Santa Cruz study area grid wells												
1	SC-01	No CDPH data	PSW	0.0	100.0	0.0	Natural	nd	nd	nd	nd	na
2	SC-02	No CDPH data	IRR	0.0	3.6	96.4	Urban	100	nd	nd	nd	na
3	SC-03	No CDPH data	PSW	0.0	60.9	39.1	Natural	71	46	61	15	na
4	SC-04	USGS data	PSW	0.0	19.0	81.0	Urban	360	155	355	200	na
5	No USGS well	No CDPH data	nd	nd	nd	nd	nd	nd	nd	nd	nd	na
6	SC-05	No CDPH data	PSW	0.0	24.2	75.8	Urban	1,700	700	1,670	970	na
6	SC-DPH-33		PSW	0.0	52.0	48.0	Natural	nd	nd	nd	nd	na
7	No USGS well	No CDPH data	nd	nd	nd	nd	nd	nd	nd	nd	nd	na
8	No USGS well	No CDPH data	nd	nd	nd	nd	nd	nd	nd	nd	nd	na
9	SC-06	USGS data	PSW	0.0	1.5	98.5	Urban	230	110	200	90	na
10	SC-07	USGS data	PSW	0.0	6.1	93.9	Urban	656	232	644	412	na
11	SC-08	USGS data	DOM	0.0	51.4	48.6	Natural	238	204	238	34	na

Table A1. Percent land use by category, land use classification, well construction information, the normalized position of selected wells along a flowpath, cell number and USGS GAMA well identification number for well data used in the Monterey Bay and Salinas Valley Basins study unit, California GAMA Priority Basin Project.—Continued

[USGS-GAMA well identification number: MB, Monterey Bay study area well;PR, Paso Robles area well;SC, Santa Cruz study area well; SV, Salinas Valley study area well. Data source: DG, use of CDPH inorganic constituent data at a USGS-grid well; DPH, a CDPH-grid well with no USGS data; FP, flow-path well; no CDPH data, no CDPH water-quality data available; no USGS well, USGS GAMA-grid well was not sampled in that cell; USGS data, USGS-GAMA data available for the grid well. Well type: DOM, domestic; IRR, irrigation; MW, monitoring; PSW, public supply. Construction information: Well depth and top and bottom of perforations measured in feet below land surface datum (LSD). Length is from the top of the uppermost perforated interval to the bottom of the well. Normalized position of well along flowpath is dimensionless; measured on a scale of 0 to 1, with 0 being the most proximal (upgradient) position and 1 being the most distal (downgradient) position along the flowpath. nd, no data; na, not available or not applicable]

Grid cell No.	USGS-GAMA well identification number	Data source	Well type	Land use within 500 meters of the well, in percent			Land-use classification	Construction information, in feet				Normalized position of well along flowpath
				Agricultural	Natural	Urban		Well depth	Depth to top of perforations	Depth to bottom of perforations	Length	
Santa Cruz study area grid wells—Continued												
12	No USGS well No CDPH data	nd	nd	nd	nd	nd	nd	nd	nd	nd	nd	na
13	SC-09	No CDPH data	PSW	0.0	99.9	0.1	Natural	240	nd	nd	nd	na
14	No USGS well No CDPH data	nd	nd	nd	nd	nd	nd	nd	nd	nd	nd	na
15	No USGS well No CDPH data	nd	nd	nd	nd	nd	nd	nd	nd	nd	nd	na
16	No USGS well No CDPH data	nd	nd	nd	nd	nd	nd	nd	nd	nd	nd	na
17	SC-10	SC-DG-10	PSW	0.2	45.1	54.6	Urban	540	380	520	140	na
18	SC-11	No CDPH data	PSW	0.0	95.9	4.1	Natural	320	255	320	65	na
18		SC-DPH-32	PSW	0.2	91.1	8.7	Natural	nd	nd	nd	nd	na
19	SC-12	No CDPH data	PSW	0.0	99.0	1.0	Natural	nd	nd	nd	nd	na
20	SC-13	No CDPH data	PSW	0.0	100.0	0.0	Natural	nd	nd	nd	nd	na
21	No USGS well No CDPH data	nd	nd	nd	nd	nd	nd	nd	nd	nd	nd	na
USGS-understanding wells												
12	MBFP-01	USGS data	PSW	0.0	45.8	54.2	Urban	1,660	970	1,650	680	0.951
29	MBFP-02	USGS data	PSW	3.2	15.3	81.4	Urban	524	360	504	144	0.820
30	MBFP-03	USGS data	PSW	0.0	4.8	95.2	Urban	342	120	340	220	0.844
30	MBMW-01	USGS data	MW	3.0	0.6	96.4	Urban	728	nd	nd	na	na
30	MBMW-02	USGS data	MW	3.0	0.6	96.4	Urban	610	nd	nd	na	na
30	MBMW-03	USGS data	MW	3.0	0.6	96.4	Urban	498	nd	nd	na	na

Table A2. Oxidation-reduction constituents, redox classification, and iron, arsenic, and chromium speciation ratios for samples from the Monterey Bay and Salinas Valley Basins study unit, California GAMA Priority Basin Project.

[Anoxic/suboxic, dissolved oxygen < 0.5 mg/L; indeterminant, insufficient data to determine redox classification; mg/L, milligram per liter; µg/L, microgram per liter; ns, not sampled; oxic, dissolved oxygen ≥ 0.5 mg/L; redox, oxidation-reduction; USGS, U.S. Geological Survey; >, greater than; ≥, greater than or equal to; <, less than; nd, not detected]

USGS-GAMA well identification number	Oxidation-reduction constituent					Redox classification	Fe(III)/ Fe(II)	As(V)/ As(III)	Cr(VI)/ Cr(III)
	Dissolved oxygen	Nitrate, as nitrogen	Manganese	Iron	Sulfate				
	Oxidation-reduction threshold value								
	≥0.5	>0.5	>50	>100	>4.0				
	Possible redox type if concentration > redox threshold value								
	O₂	NO₃	Mn	Fe	SO₄				
	Analysis reporting level and associated units								
	0.1	0.06	0.18	5.0	0.18				
	mg/L	mg/L	µg/L	µg/L	mg/L				
Monterey Bay study area									
MB-01	8.6	2.4	nd	nd	85.4	Oxic	ns	ns	9.0
MB-02	2.2	5.5	14.0	nd	101	Oxic	ns	ns	nd
MB-03	1.9	ns	ns	ns	ns	Oxic	ns	ns	16.0
MB-04	0.1	nd	13.3	48.0	61.9	Anoxic/suboxic	4.0	nd	nd
MB-05	7.8	1.3	nd	nd	19.0	Oxic	ns	ns	7.3
MB-06	1.2	nd	110	nd	29.0	Mixed	ns	ns	nd
MB-07	4.3	nd	nd	220	59.0	Mixed	ns	ns	8.0
MB-08	2.5	ns	ns	ns	ns	Oxic	ns	ns	>10
MB-09	2.9	0.09	0.6	18.0	31.6	Oxic	nd	nd	3.3
MB-10	4.3	3.4	ns	ns	ns	Oxic	ns	ns	>10
MB-11	1.3	0.90	nd	105	14.0	Mixed	ns	ns	>10
MB-12	0.2	nd	5.5	11.0	46.4	Anoxic/suboxic	0.1	>10	>10
MB-13	3.9	ns	ns	ns	ns	Oxic	ns	ns	>10
MB-14	4.1	0.8	28.0	nd	93.8	Oxic	ns	ns	7.0
MB-15	0.6	ns	ns	ns	ns	Oxic	ns	ns	>10
MB-16	4.1	ns	ns	ns	ns	Oxic	ns	ns	4.0
MB-17	1.8	0.45	nd	nd	35.0	Oxic	ns	ns	6.0
MB-18	3.5	0.7	nd	nd	14.4	Oxic	nd	>10	>10
MB-19	2.9	0.8	20.0	200	13.0	Mixed	ns	ns	>10
MB-20	0.9	1.6	1.6	nd	54.3	Oxic	nd	nd	7.0
MB-21	0.2	ns	ns	ns	ns	Anoxic/suboxic	ns	ns	27.0
MB-22	0.1	nd	60.8	90.0	125	Anoxic/suboxic	0.1	nd	>10
MB-23	2.7	nd	ns	ns	ns	Oxic	ns	ns	5.3
MB-24	7.5	3.5	nd	nd	5.0	Oxic	ns	ns	>10
MB-25	4.5	0.7	nd	nd	13.0	Oxic	ns	ns	>10
MB-26	2.7	1.0	0.4	nd	144	Oxic	ns	ns	>10
MB-27	ns	1.5	155	41.2	79.0	Mixed	ns	ns	6.0
MB-28	ns	1.8	3.9	50.2	62.8	Indeterminate	ns	ns	>10
MB-29	3.5	4.7	3.2	nd	445	Oxic	nd	>10	>10
MB-30	4.0	3.9	1.3	9.0	88.1	Oxic	7.0	>10	3.0
MB-31	5.2	6.7	9.6	96.2	29.4	Oxic	ns	ns	nd
MB-32	8.4	ns	ns	ns	ns	Oxic	ns	ns	>10
MB-33	1.4	1.5	208	87.0	6.8	Mixed	0.4	1.0	2.0
MB-34	0.2	nd	133	1,470	34.0	Anoxic/suboxic	ns	ns	nd
MB-35	nd	nd	2,410	35.0	216	Anoxic/suboxic	0.1	nd	2.5
MB-36	0.4	0.5	nd	131	17.0	Mixed	ns	ns	nd

Table A2. Oxidation-reduction constituents, redox classification, and iron, arsenic, and chromium speciation ratios for samples from the Monterey Bay and Salinas Valley Basins study unit, California GAMA Priority Basin Project.—Continued

[Anoxic/suboxic, dissolved oxygen < 0.5 mg/L; oxic, dissolved oxygen ≥ 0.5 mg/L; indeterminant, insufficient data to determine redox classification; mg/L, milligram per liter; µg/L, microgram per liter; ns, not sampled; redox, oxidation-reduction; USGS, U.S. Geological Survey; >, greater than; ≥, greater than or equal to; <, less than; nd, not detected]

USGS-GAMA well identification number	Oxidation-reduction constituent					Redox classification	Fe(III)/ Fe(II)	As(V)/ As(III)	Cr(VI)/ Cr(III)
	Dissolved oxygen	Nitrate, as nitrogen	Manganese	Iron	Sulfate				
	Oxidation-reduction threshold value								
	≥0.5	>0.5	>50	>100	>4.0				
	Possible redox type if concentration > redox threshold value								
	O_2	NO_3	Mn	Fe	SO_4				
	Analysis reporting level and associated units								
	0.1	0.06	0.18	5.0	0.18				
	mg/L	mg/L	µg/L	µg/L	mg/L				
Monterey Bay study area—Continued									
MB-37	5.4	1.6	nd	nd	15.2	Oxic	>10	>10	>10
MB-38	4.3	4.4	nd	nd	90.0	Oxic	ns	ns	>10
MB-39	ns	ns	ns	ns	ns	Indeterminate	ns	ns	nc
MB-40	4.3	1.2	0.1	nd	133	Oxic	4.3	>10	>10
MB-41	0.7	ns	ns	ns	ns	Oxic	ns	ns	>10
MB-42	9.4	ns	ns	ns	ns	Oxic	ns	ns	nd
MB-43	1.8	ns	ns	ns	ns	Oxic	ns	ns	6.0
MB-44	5.5	37.8	0.6	4.0	387	Oxic	>10	>10	11.0
MB-45	1.3	nd	170	2,830.0	102	Mixed	0.004	>10	2.0
MB-46	0.7	ns	ns	ns	ns	Oxic	ns	ns	>10
MB-47	2.8	0.6	0.8	10.0	92.6	Oxic	2.3	>10	>10
MB-48	12.0	2.7	ns	ns	ns	Oxic	ns	ns	17.0
MB-DPH-50	1.9	nd	nd	nd	1,700	Oxic	ns	ns	ns
MB-DPH-51	2.5	nd	nd	nd	36.0	Oxic	ns	ns	ns
MB-DPH-52	3.9	nd	nd	nd	78.0	Oxic	ns	ns	ns
MB-DPH-54	ns	0.9	ns	ns	ns	Indeterminate	ns	ns	ns
MB-DPH-55	0.7	1.1	nd	nd	126	Oxic	ns	ns	ns
MB-DPH-56	9.4	0.2	nd	nd	82.0	Oxic	ns	ns	ns
MB-DPH-57	0.7	18.5	nd	412	307	Mixed	ns	ns	ns
MB-DPH-68	8.4	4.1	nd	nd	5.2	Oxic	ns	ns	ns
MB-DPH-71	1.8	20.3	nd	nd	78.0	Oxic	ns	ns	ns
MBFP-01	0.3	ns	ns	ns	ns	Anoxic/suboxic	ns	ns	nd
MBFP-02	1.4	11.6	nd	3.0	223	Oxic	nd	>10	>10
MBFP-03	1.3	9.3	0.3	23.0	161	Oxic	2.3	>10	7.0
MBMW-01	2.2	0.8	0.4	nd	32.7	Oxic	>10	>10	7.0
MBMW-02	4.6	1.0	0.8	nd	30.9	Oxic	>10	>10	>10
MBMW-03	4	1	1.0	nd	26.3	Oxic	>10	>10	11.0
Paso Robles study area									
PR-01	4.0	2.2	0.2	4.0	126	Oxic	0.5	>10	nd
PR-02	2.0	10.4	nd	nd	180	Oxic	ns	ns	5.0
PR-03	1.7	nd	60.0	nd	130	Mixed	ns	ns	2.0
PR-04	1.1	ns	ns	ns	ns	Oxic	ns	ns	5.0
PR-05	0.2	0.5	nd	nd	70	Anoxic/suboxic	ns	ns	5.0
PR-06	6.2	6.8	nd	nd	28	Oxic	ns	ns	>10
PR-07	ns	4.7	nd	nd	23	Indeterminate	ns	ns	>10

Table A2. Oxidation-reduction constituents, redox classification, and iron, arsenic, and chromium speciation ratios for samples from the Monterey Bay and Salinas Valley Basins study unit, California GAMA Priority Basin Project.—Continued

[Anoxic/suboxic, dissolved oxygen < 0.5 mg/L; oxic, dissolved oxygen ≥ 0.5 mg/L; indeterminant, insufficient data to determine redox classification; mg/L, milligram per liter; µg/L, microgram per liter; ns, not sampled; redox, oxidation-reduction; USGS, U.S. Geological Survey; >, greater than; ≥, greater than or equal to; <, less than; nd, not detected]

USGS-GAMA well identification number	Oxidation-reduction constituent					Redox classification	Fe(III)/ Fe(II)	As(V)/ As(III)	Cr(VI)/ Cr(III)
	Dissolved oxygen	Nitrate, as nitrogen	Manganese	Iron	Sulfate				
	Oxidation-reduction threshold value								
	≥0.5	>0.5	>50	>100	>4.0				
	Possible redox type if concentration > redox threshold value								
	O₂	NO₃	Mn	Fe	SO₄				
	Analysis reporting level and associated units								
	0.1	0.06	0.18	5.0	0.18				
	mg/L	mg/L	µg/L	µg/L	mg/L				
Paso Robles study area—Continued									
PR-08	0.8	3.6	1.0	10.0	173	Oxic	0.4	>10	>10
PR-09	ns	0.1	nd	100	260	Anoxic/suboxic	ns	ns	>10
PR-10	0.1	0.4	78.4	185	563	Anoxic/suboxic	0.2	>10	nd
PR-11	ns	ns	ns	ns	ns	Indeterminate	ns	ns	>10
PR-DPH-21	1.1	nd	nd	nd	110	Oxic	ns	ns	ns
PR-DPH-22	ns	0.2	nd	nd	23	Indeterminate	ns	ns	ns
PR-DPH-23	ns	3.4	nd	nd	63	Indeterminate	ns	ns	ns
Santa Cruz study area									
SC-01	8.4	ns	ns	ns	ns	Oxic	ns	ns	>10
SC-02	0.2	ns	ns	ns	ns	Anoxic/suboxic	ns	ns	5.0
SC-03	2.8	ns	ns	ns	ns	Oxic	ns	ns	2.0
SC-04	0.4	nd	13.8	37.0	180	Anoxic/suboxic	0.03	nd	2.0
SC-05	1.8	ns	ns	ns	ns	Oxic	ns	ns	1.8
SC-06	0.2	nd	197	657	102	Anoxic/suboxic	0.03	nd	>10
SC-07	0.3	nd	38.8	92.0	38.9	Anoxic/suboxic	0.4	nd	>10
SC-08	0.2	nd	63.0	218	26.6	Anoxic/suboxic	0.09	>10	3.0
SC-09	6.6	ns	ns	ns	ns	Oxic	ns	ns	5.3
SC-10	5.5	ns	ns	ns	ns	Oxic	ns	ns	>10
SC-11	0.1	ns	ns	ns	ns	Anoxic/suboxic	ns	ns	>10
SC-12	0.2	ns	ns	ns	ns	Anoxic/suboxic	ns	ns	7.0
SC-13	3.0	ns	ns	ns	ns	Oxic	ns	ns	3.4
Salinas Valley study area									
SV-01	0.1	nd	34.5	508	449	Anoxic/suboxic	0.02	0.3	>10
SV-02	0.1	0.3	5.1	nd	78.4	Anoxic/suboxic	0.3	>10	>10
SV-03	0.6	0.0	536	33.0	272	Mixed	0.6	>10	7.0
SV-04	ns	ns	ns	ns	ns	Indeterminate	ns	ns	>10
SV-05	nd	10.6	nd	nd	113	Anoxic/suboxic	ns	ns	4.0
SV-06	ns	0.7	nd	78.0	81.0	Indeterminate	ns	ns	8.0
SV-07	0.1	1.0	0.2	6.0	79.2	Anoxic/suboxic	nd	>10	>10
SV-08	3.5	ns	ns	ns	ns	Oxic	ns	ns	4.4
SV-09	4.7	ns	ns	ns	ns	Oxic	ns	ns	5.2
SV-10	2.2	ns	ns	ns	ns	Oxic	ns	ns	4.2
SV-11	8.0	1.9	0.2	nd	80.8	Oxic	>10	nd	3.0
SV-12	ns	ns	ns	ns	ns	Indeterminate	ns	ns	2.5
SV-13	10.4	ns	ns	ns	ns	Oxic	ns	ns	3.0

Table A2. Oxidation-reduction constituents, redox classification, and iron, arsenic, and chromium speciation ratios for samples from the Monterey Bay and Salinas Valley Basins study unit, California GAMA Priority Basin Project.—Continued

[Anoxic/suboxic, dissolved oxygen < 0.5 mg/L; oxic, dissolved oxygen ≥ 0.5 mg/L; indeterminant, insufficient data to determine redox classification; mg/L, milligram per liter; μg/L, microgram per liter; ns, not sampled; redox, oxidation-reduction; USGS, U.S. Geological Survey; >, greater than; ≥, greater than or equal to; <, less than; nd, not detected]

USGS-GAMA well identification number	Oxidation-reduction constituent					Redox classification	Fe(III)/ Fe(II)	As(V)/ As(III)	Cr(VI)/ Cr(III)
	Dissolved oxygen	Nitrate, as nitrogen	Manganese	Iron	Sulfate				
	Oxidation-reduction threshold value								
	≥0.5	>0.5	>50	>100	>4.0				
	Possible redox type if concentration > redox threshold value								
	O_2	NO_3	Mn	Fe	SO_4				
	Analysis reporting level and associated units								
	0.1	0.06	0.18	5.0	0.18				
	mg/L	mg/L	μg/L	μg/L	mg/L				
Salinas Valley study area—Continued									
SV-14	ns	ns	ns	ns	ns	Indeterminate	ns	ns	5.0
SV-15	ns	ns	ns	ns	ns	Indeterminate	ns	ns	>10
SV-16	7.7	1.1	ns	ns	ns	Oxic	ns	ns	>10
SV-17	4.5	ns	ns	ns	ns	Oxic	ns	ns	>10
SV-18	0.3	0.2	19.3	24.0	111	Anoxic/suboxic	1.3	>10	3.0
SV-19	2.5	1.0	2.7	7.0	136	Oxic	ns	ns	>10
SV-DPH-40	ns	1.4	ns	ns	ns	Indeterminate	ns	ns	ns
SV-DPH-42	ns	9.0	23.0	123	151	Mixed	ns	ns	ns
SV-DPH-44	ns	3.4	nd	nd	173	Indeterminate	ns	ns	ns
SV-DPH-45	4.7	10.4	nd	nd	120	Oxic	ns	ns	ns

Table A3. Noble-gas-based recharge temperature, tritium, terrigenic helium, percent modern carbon, and age classification of samples, Monterey Bay and Salinas Valley Basins study unit, California GAMA Priority Basin Project, July–October 2005.

[Modern, recharged after 1953; mixed, modern and pre-modern water; pre-modern, recharged before 1953; °C, degrees Celsius; ns, not sampled; <, less than; USGS, U.S. Geological Survey]

USGS-GAMA well identification number	Noble-gas-based recharge temperature, in °C	Tritium, in tritium units	Terrigenic helium, percentage of total helium	Percent modern carbon	Age classification	USGS-GAMA well identification number	Noble-gas-based recharge temperature, in °C	Tritium, in tritium units	Terrigenic helium, percentage of total helium	Percent modern carbon	Age classification
MB-01	16.1	0.3	25.3	ns	Pre-modern	MBFP-02	14.2	3.1	5.2	93.4	Mixed
MB-02	16.8	1.4	62.2	ns	Mixed	MBFP-03	12.8	1.2	0.5	90.3	Modern
MB-03	10.8	<1	89.1	ns	Pre-modern	MBMW-01	12.1	<1	93.4	19.8	Pre-modern
MB-04	6.8	0.7	71.8	5.7	Pre-modern	MBMW-02	13.1	<1	86.7	43.5	Pre-modern
MB-05	15.5	1.2	0.0	ns	Modern	MBMW-03	15.9	<1	52.0	72.7	Pre-modern
MB-06	11.1	0.5	0.0	ns	Mixed	PR-01	15.1	2.2	4.9	91.0	Modern
MB-07	7.6	0.2	34.6	ns	Pre-modern	PR-02	15.6	1.4	86.4	ns	Mixed
MB-08	10.0	1.2	5.4	ns	Mixed	PR-03	12.3	0.1	88.6	ns	Pre-modern
MB-09	13.0	0.5	0.0	73.4	Mixed	PR-04	14.1	<1	90.0	ns	Pre-modern
MB-10	13.6	0.2	2.1	ns	Mixed	PR-05	13.2	0.2	90.6	ns	Pre-modern
MB-11	13.2	0.3	92.3	ns	Pre-modern	PR-06	20.5	0.5	9.3	ns	Pre-modern
MB-12	12.9	<1	77.5	6.3	Pre-modern	PR-07	18.5	<1	25.1	ns	Pre-modern
MB-13	16.0	0.2	18.4	ns	Pre-modern	PR-08	14.4	0.2	38.1	47.6	Pre-modern
MB-14	13.9	0.2	10.8	ns	Pre-modern	PR-09	14.5	<1	95.3	ns	Pre-modern
MB-15	15.1	0.7	0.0	ns	Mixed	PR-10	15.2	<1	74.4	67.5	Pre-modern
MB-16	13.9	0.2	21.1	ns	Pre-modern	PR-11	14.0	<1	14.1	ns	Pre-modern
MB-17	12.8	<1	95.3	ns	Pre-modern	SV-01	14.7	<1	95.1	8.9	Pre-modern
MB-18	16.3	<1	28.0	56.1	Pre-modern	SV-02	15.5	2.1	0.0	100.6	Modern
MB-19	15.3	0.1	0.0	ns	Mixed	SV-03	16.8	2.1	87.7	99.5	Mixed
MB-20	14.1	0.8	0.0	77.0	Mixed	SV-04	17.8	0.7	6.1	ns	Pre-modern
MB-21	11.8	<1	0.0	ns	Mixed	SV-05	16.0	2.0	95.0	ns	Mixed
MB-22	14.6	0.3	0.0	80.5	Mixed	SV-06	17.3	2.1	0.0	ns	Modern
MB-23	14.7	<1	7.2	ns	Pre-modern	SV-07	17.2	1.9	0.0	100.2	Modern
MB-24	16.6	0.3	0.8	ns	Mixed	SV-08	16.6	2.3	26.2	ns	Mixed
MB-25	13.1	0.1	88.8	ns	Pre-modern	SV-09	18.2	2.0	0.0	ns	Modern
MB-26	13.2	0.4	43.7	ns	Pre-modern	SV-10	18.2	2.5	0.0	ns	Modern
MB-27	12.1	0.2	68.7	ns	Pre-modern	SV-11	13.4	2.7	3.2	89.8	Mixed
MB-28	13.0	0.5	6.9	ns	Pre-modern	SV-12	14.8	2.5	0.0	ns	Modern
MB-29	17.0	2.8	0.0	96.4	Modern	SV-13	14.8	1.6	0.0	ns	Modern
MB-30	12.4	0.1	44.2	79.3	Pre-modern	SV-14	13.0	1.8	22.3	ns	Mixed
MB-31	12.3	2.1	0.0	ns	Modern	SV-15	17.0	2.2	0.0	ns	Modern
MB-32	17.3	0.4	0.0	ns	Mixed	SV-16	14.6	0.0	0.0	ns	Mixed
MB-33	15.2	0.5	76.4	54.0	Pre-modern	SV-17	13.0	1.1	0.0	ns	Modern
MB-34	12.2	0.3	28.5	ns	Pre-modern	SV-18	ns	<1	0.0	16.3	Mixed
MB-35	17.3	2.5	0.0	97.3	Modern	SV-19	13.8	0.1	62.1	33.8	Pre-modern
MB-36	11.6	0.5	85.5	ns	Pre-modern	SC-01	ns	2.4	0.0	ns	Modern
MB-37	16.1	<1	96.4	57.5	Pre-modern	SC-02	ns	2.4	0.0	ns	Modern
MB-38	12.5	0.6	49.1	ns	Pre-modern	SC-03	15.6	2.1	1.0	ns	Modern
MB-39	13.1	0.3	18.2	ns	Pre-modern	SC-04	14.0	1.7	58.2	65.9	Mixed
MB-40	10.4	0.5	19.2	73.0	Pre-modern	SC-05	ns	<1	0.0	ns	Mixed
MB-41	16.8	3.7	0.0	ns	Modern	SC-06	14.9	1.4	0.0	71.6	Mixed
MB-42	ns	2.0	0.0	ns	Modern	SC-07	14.2	0.4	60.7	31.5	Pre-modern
MB-43	15.4	3.2	0.0	ns	Modern	SC-08	17.1	<1	0.0	57.5	Mixed
MB-44	16.7	2.3	0.0	105.7	Modern	SC-09	15.9	1.4	0.0	ns	Modern
MB-45	16.9	2.2	3.9	92.1	Modern	SC-10	9.4	-0.1	22.4	ns	Pre-modern
MB-46	17.0	2.2	0.0	ns	Modern	SC-11	ns	<1	0.0	ns	Mixed
MB-47	12.2	<1	54.8	32.5	Pre-modern	SC-12	12.6	<1	0.0	ns	Mixed
MB-48	15.8	1.4	3.3	ns	Modern	SC-13	12.5	4.3	0.0	ns	Modern
MBFP-01	10.4	0.2	95.2	ns	Pre-modern						

Appendix B. Use of Data From the California Department of Public Health (CDPH) Database

For the MS study unit, the historical CDPH database contains more than 502,000 records distributed across more than 850 wells, requiring targeted retrievals to manageably use the data to assess water quality. The following paragraphs summarize the selection process for wells and data from the CDPH database for use in the grid-based *status assessment*.

The strategy used to select CDPH inorganic data for a single well in each cell where the USGS did not obtain a sample for analysis for inorganic constituents involved prioritizing data from different sources. The first choice was to select CDPH data for the grid well sampled by the USGS (figs. B3, B4) for other constituents, provided the CDPH data met quality-control criteria. Cation/anion balance was used as the quality-control assessment metric. Because water is electrically neutral and must have a balance between positive (cations) and negative (anions) electrically charged dissolved species, the cation/anion balance commonly is used as a quality-assurance criterion for water sample analysis (Hem, 1970). An imbalance equal to or greater than 10 percent may indicate uncertainty in the quality of the data or that data were missing for one or more constituents necessary to achieve balance. The most recent CDPH data from the well were evaluated to determine whether the cation/anion imbalance was less than 10 percent; if so, the CDPH inorganic data for the well were selected for use as grid-well data (USGS-grid well with CDPH inorganic data [figs. B1, B2]). It was assumed that if analyses met quality-control criteria—cation/anion balance—for major and minor elements, then analyses at these wells for trace elements, nutrients, and radiochemical constituents also would be of acceptable quality. This approach resulted in the selection of inorganic data from the CDPH database for 11 USGS-grid wells. To identify the USGS-wells that incorporated CDPH inorganic data, a well ID was created that added "DG" to the GAMA ID for these wells (for example, MB-01 with CDPH data was assigned the well identification MB-DG-01; table A1).

If the first step did not yield CDPH inorganic data for the USGS-grid well, the second step was to search the CDPH database to identify the highest ranked well with a cation/anion imbalance less than 10 percent in each grid cell. This step resulted in selecting CDPH inorganic data for non-USGS-sampled wells for 11 grid cells. These 11 CDPH-grid wells were not co-located with their cell's respective USGS-grid well. To identify these new CDPH grid wells, a well ID was created that added "DPH" after the study unit prefix and then a sequential number starting after the last GAMA ID for the study area (for example, CDPH-grid well MB-DPH-50, table A1). If no wells in a grid cell met the cation/anion balance criteria or if there was insufficient data to evaluate charge balance, the third choice for the CDPH-grid well was to select the highest randomly ranked CDPH well with any of the needed inorganic data. This resulted in selecting CDPH inorganic data for 23 USGS-grid wells and 16 additional wells. If the well was a USGS-grid well, then a well ID was created that added "DG" to the GAMA ID (for example, MB-DG-01), or if the well was a new CHDP-grid well, then "DPH" was added after the study unit prefix and then a sequential number starting after the last GAMA ID for the study area (for example, CDPH-grid well SC-DPH-32). In some cases, to achieve one value for each constituent per cell, it was necessary to select an additional well in a cell for data: hence, some cells have multiple CDPH wells.

The result of these steps was one grid well per cell having data from the USGS database, the CDPH database, or both database. Inorganic data from the CDPH database were used for 63 grid wells. Data were available for 47 grid wells for nitrate plus nitrite and for 0 to 44 wells for most other inorganic constituents (table 2). In combination with USGS-grid well inorganic data (29 wells), inorganic data was available for 90 of the 116 grid cells. Estimates of aquifer-scale proportion for constituents based on a smaller number of wells are subject to a larger error associated with the 90 percent confidence intervals (on the basis of Jeffreys interval for the binomial distribution).

Differences in constituent reporting levels associated with USGS and CDPH data did not affect analysis of high or moderate relative-concentrations because concentrations greater than one-half of water-quality benchmarks were substantially higher than the reporting levels. Several types of comparisons between USGS-collected and CDPH data are described in appendix E.

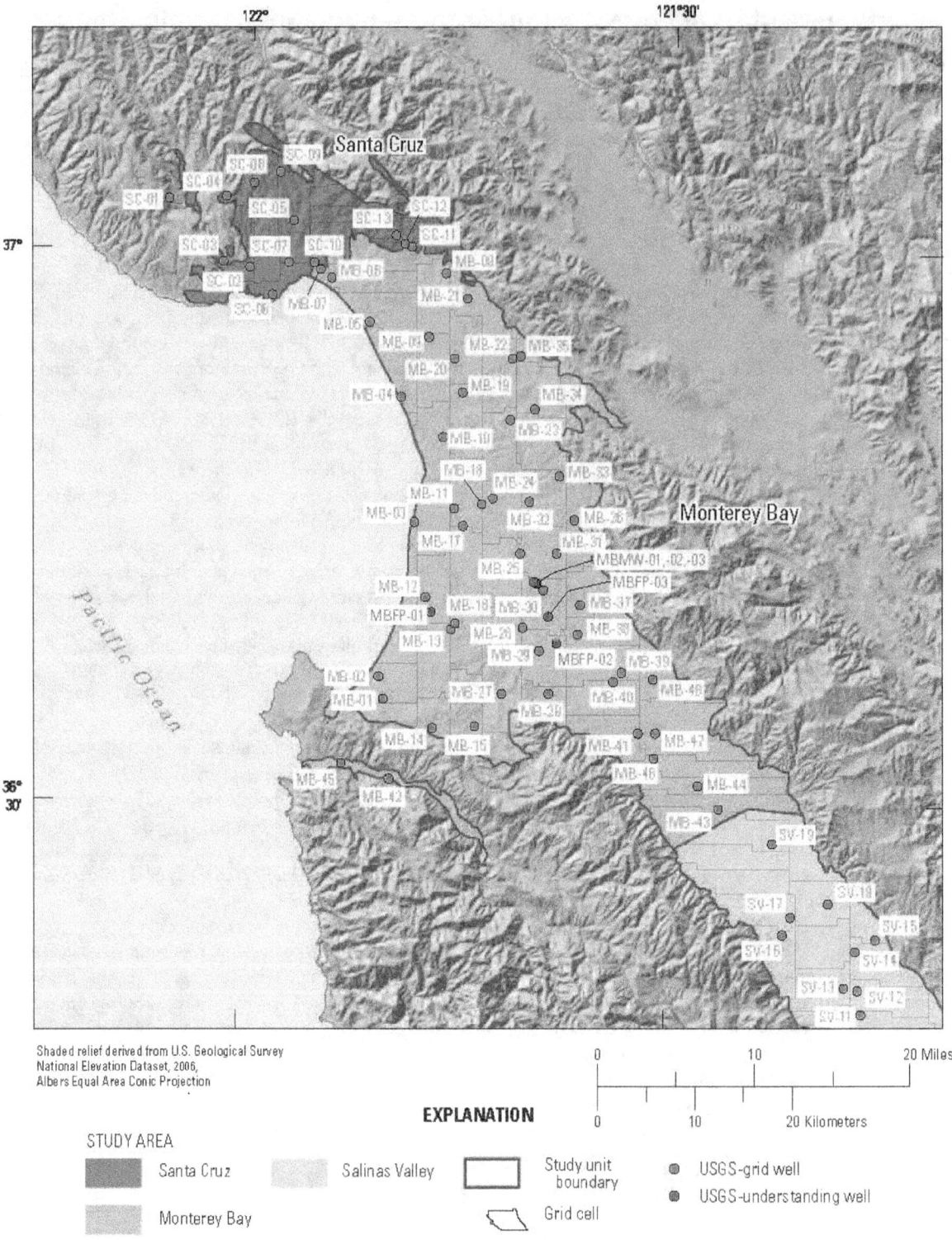

Figure B1. Identifiers and locations of USGS-grid and USGS-understanding wells sampled in the Santa Cruz and Monterey Bay study areas during July–November 2005, Monterey Bay and Salinas Valley Basins study unit, California GAMA Priority Basin Project.

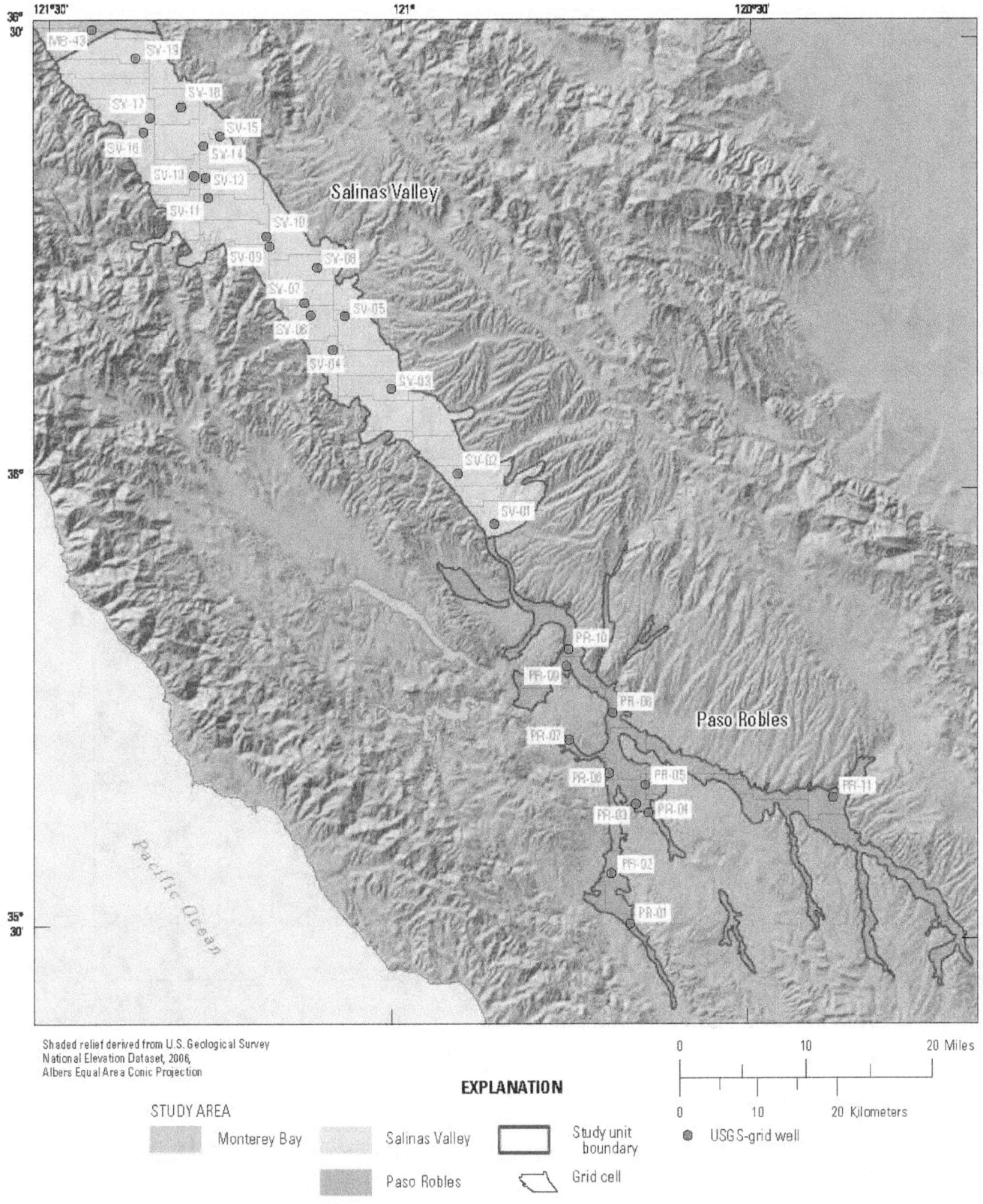

Figure B2. Identifiers and locations of USGS-grid wells sampled in the Salinas Valley and Paso Robles study areas during July–November 2005, Monterey Bay and Salinas Valley Basins study unit, California GAMA Priority Basin Project.

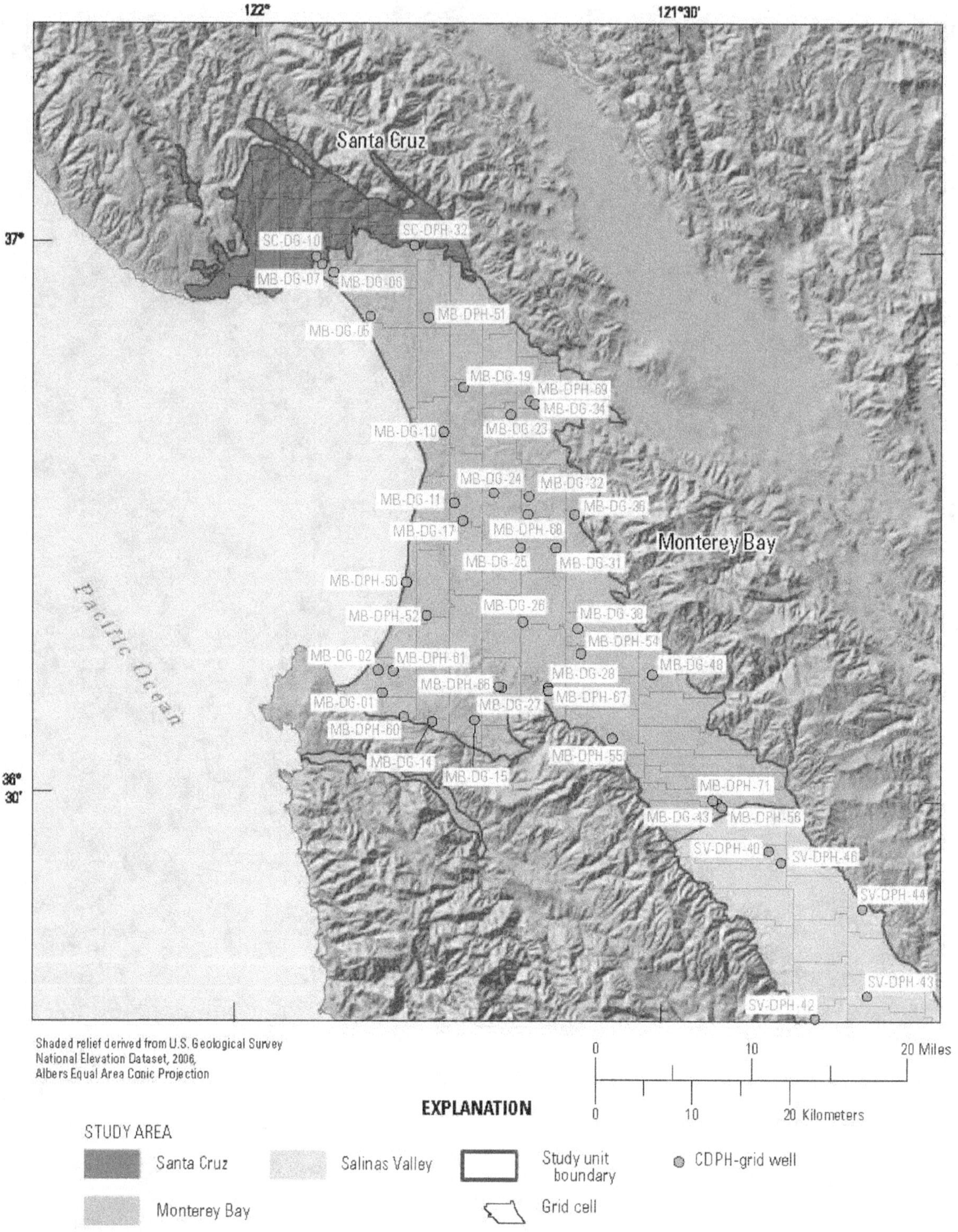

Figure B3. Identifiers and locations of CDPH-grid wells, Santa Cruz and Monterey Bay study areas, Monterey Bay and Salinas Valley Basins study unit, California GAMA Priority Basin Project.

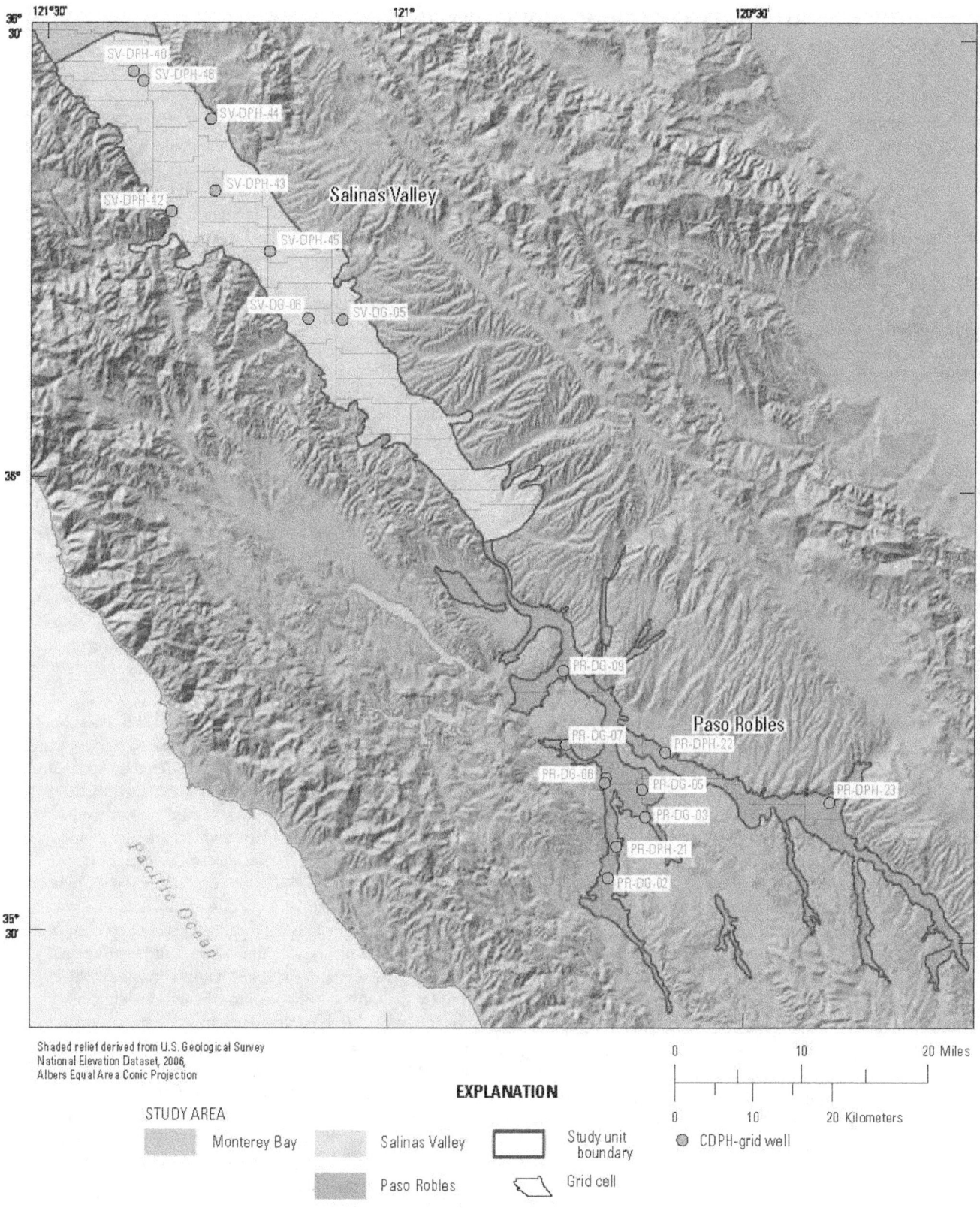

Figure B4.   Identifiers and locations of CDPH-grid wells, Salinas Valley and Paso Robles study areas, Monterey Bay and Salinas Valley Basins study unit, California GAMA Priority Basin Project.

Appendix C. Calculation of Aquifer-Scale Proportions

The *status assessment* is intended to characterize the quality of groundwater resources in the primary aquifers of the MS study unit. The primary aquifers are defined by the perforated depth intervals of the wells listed in the CDPH database. The use of the term "primary aquifers" does not imply that there exists a discrete aquifer unit. In most groundwater basins, municipal and community supply wells generally are perforated at greater depths than domestic wells. Thus, because domestic wells are not listed in the CDPH database, the primary aquifers generally corresponds to the portion of the aquifer system tapped by municipal and community supply wells. A majority of the wells used in the *status assessment* are listed in the CDPH database and are therefore classified as municipal and community drinking-water supply wells. However, to the extent that domestic wells are perforated over the same depth intervals as the CDPH wells, the assessments presented in this report also may be applicable to the portions of the aquifer systems used for domestic drinking-water supplies.

Two statistical approaches, grid-based and spatially weighted, were selected to evaluate the aquifer-scale proportions of the primary aquifers in the MS study unit with high, moderate, or low relative-concentrations of constituents relative to benchmarks(Belitz and others, 2010). The grid-based and spatially weighted estimations of aquifer-scale proportions, based on a spatially distributed grid cell network across the MS study unit, are intended to characterize the water quality of the primary aquifers, or at depths from which drinking water is usually drawn. These approaches assign weights to wells based on a single well per cell (grid-based) or the number of wells per cells (spatially weighted).

Raw detection frequencies, derived from the percentage of the total number of wells with high or moderate relative-concentrations, also were calculated for individual constituents, but were not used for estimating aquifer-scale proportion because this method creates spatial bias towards regions with large numbers of wells.

1. Grid-based. One well in each grid cell, a "grid well," was randomly selected to represent the primary aquifers (Belitz and others, 2010). Most grid wells sampled for the MS study were USGS-grid wells. However, data for all constituents were not available for some USGS-grid wells and additional data for CDPH-grid wells were selected to provide data for grid cells with no USGS-grid wells. The relative-concentration for each constituent (concentration relative to its benchmark) was then evaluated for each grid well. The proportion of the primary aquifers with high relative-concentrations was calculated by dividing the number of cells with concentrations greater than the benchmark (relative-concentration greater than 1) by the total number of grid wells in the MS study unit. Proportions containing moderate relative-concentrations were calculated similarly. Confidence intervals for grid-based aquifer proportions were computed using the Jeffreys interval for the binomial distribution (Brown and others, 2001). The grid-based estimate is spatially unbiased. However, the grid-based approach may not identify constituents that exist at high concentrations in small proportions of the primary aquifers.

2. Spatially weighted. The spatially weighted approach relied on USGS-grid well data collected from July–October 2005, and CDPH data from July 17, 2002–July 18, 2005 (most recent analyses per well for all wells within each grid cell), and USGS-understanding public-supply well data. However, instead of data from only one well per grid cell, the spatially weighted approach uses all wells in each cell to calculate the high, moderate, and low relative-concentrations for the cell. The high, moderate, and low aquifer-scale proportions are then calculated from the percentage of cells with high, moderate, or low relative-concentrations (Isaaks and Srivastava, 1989). The resulting proportions are spatially unbiased (Isaaks and Srivastava, 1989). Confidence intervals for spatially weighted estimates of aquifer-scale proportion are not described in this report.

The raw detection frequency approach merely is the percentage (frequency) of wells within the MS study unit with high relative-concentrations. It was calculated by considering all of the available data from July 17, 2002–July 18, 2005, for the CDPH well data (the most recent analysis per well for all wells), the USGS-grid well data, and USGS-understanding wells. However, this approach is not spatially unbiased because the CDPH and USGS-understanding wells are not uniformly distributed. Consequently, high values (or low values) for wells clustered in a particular area represent a small part of the primary aquifers, and could be given a disproportionately high (or low) weight compared to that given by spatially unbiased approaches. Raw detection frequencies of high relative-concentrations are provided to identify constituents for discussion in this report (table 4), but were not used to assess aquifer-scale proportions. For calculation of high aquifer-scale proportion for a class of constituents, cells were considered high if values for any of the constituents in that class were high. Cells were considered moderate if values for any of the constituents were moderate, but no values were high.

Appendix D. Calculating Total Dissolved Solids

Specific conductance, an electrical measure of total dissolved solids (TDS), was available for 96 USGS-grid and understanding wells, whereas TDS was only measured directly as residue on evaporation for 34 of these wells. For wells with no measured TDS, TDS was calculated from specific conductance (SC) values using a linear regression equation (TDS = 0.68*SC −17; coefficient of determination, R^2 = 0.97). Four SC values were omitted from the calculation because the values were significantly different from the other values plotted.

Appendix E. Comparison of California Department of Public Health and U.S. Geological Survey-GAMA Data

CDPH and USGS-GAMA data were compared to assess the validity of combining data from these different sources. Because laboratory reporting levels for most organic constituents and trace elements were substantially lower for USGS-GAMA data than for CDPH data (table 2), only relatively high concentrations of constituents could be compared, and as a result, there were insufficient data from which to evaluate agreement between CDPH and USGS-GAMA data.. However, concentrations of inorganic constituents (calcium, magnesium, sodium, chloride, sulfate, arsenic, and TDS), which generally are prevalent at concentrations substantially greater than reporting levels, were compared for each well using data from both sources. The USGS and CDPH databases contained data for major ion, trace element, or TDS for 8 to 19 wells. Wilcoxon signed rank tests of paired analyses for these eight constituents indicated no significant differences between USGS-GAMA and CDPH

data for these constituents. Although differences between the paired datasets occurred for some wells, most sample pairs plotted close to a 1:1 line (fig. E1). These plots indicated that the GAMA and CDPH inorganic data were comparable.

Major-ion data for grid wells with sufficient data (USGS and CDPH data) were plotted on a trilinear diagram (Piper, 1944) along with all CDPH major-ion data to determine whether the groundwater types in grid wells were similar to groundwater types observed historically in the study unit. Trilinear diagrams show the relative abundance of major cations and anions (on a charge equivalent basis) as a percentage of the total ion content of the water (fig. E2). Trilinear diagrams often are used to define groundwater type (Hem, 1970). All cation/anion data in the CDPH database with a cation/anion balance less than 10 percent were retrieved and plotted on the trilinear diagram for comparison with USGS- and CDPH-grid well data.

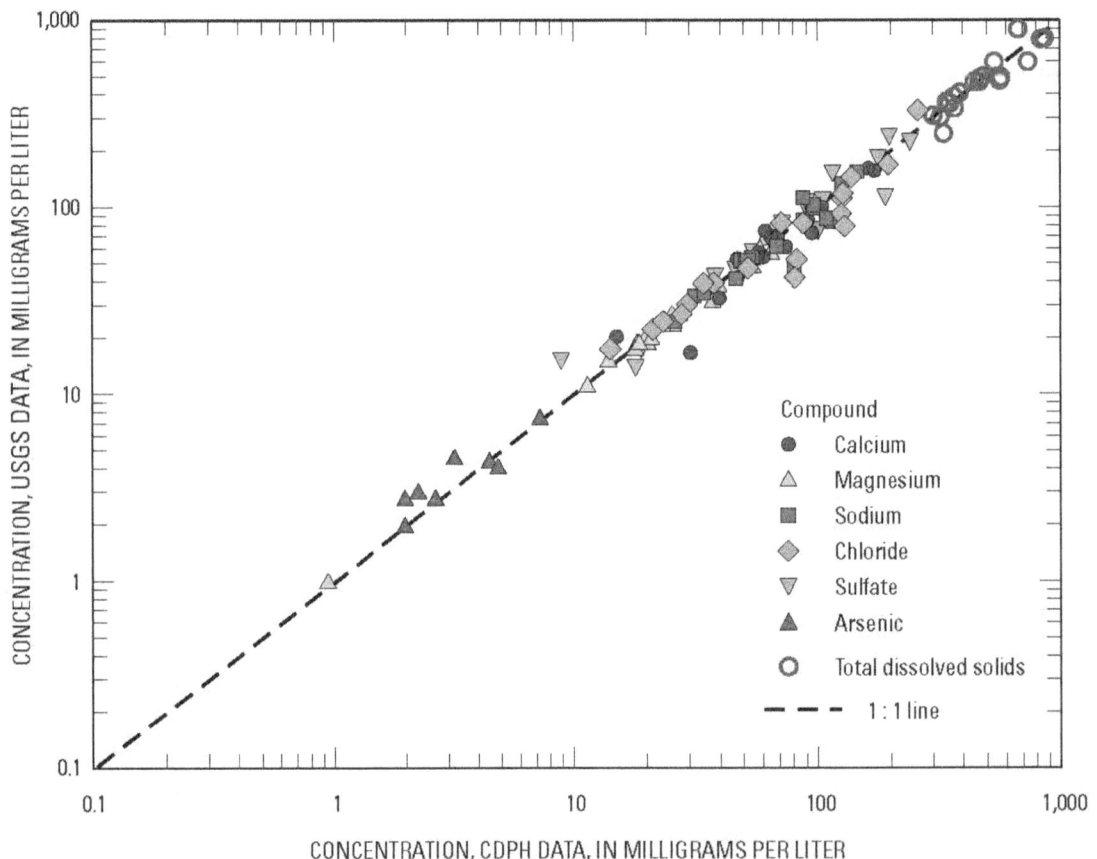

Figure E1.   Paired inorganic-constituent concentrations from wells sampled by the Groundwater Ambient Monitoring Assessment (GAMA) Program from July to October 2005 and California Department of Public Health database for the same wells from the period (July 17, 2002–July 18, 2005), Monterey Bay and Salinas Valley Basins study unit, California GAMA Priority Basin Project.

The ranges of water types for USGS-grid wells and other wells from the historical CDPH database were similar (fig. E2). In most water samples from wells, no single cation accounted for more than 10 percent of the total cations, and bicarbonate accounted for more than 10 percent of the total anions. Waters in these wells are described as *mixed cation-bicarbonate* type waters. Many wells also contained *mixed cation-mixed anion* type waters, indicating that no single cation and no single anion accounted for more than 10 percent

of the total. Waters in a minority of wells were classified as *sodium-chloride* type waters, indicating that sodium and chloride accounted for more than 10 percent of the total cations and anions, respectively.

The determination that the range of relative abundance of major cations and anions in grid wells (42 wells) is similar to the range of those in all CDPH wells (114 wells) indicates that the grid wells represent the types of water present in the MS study unit.

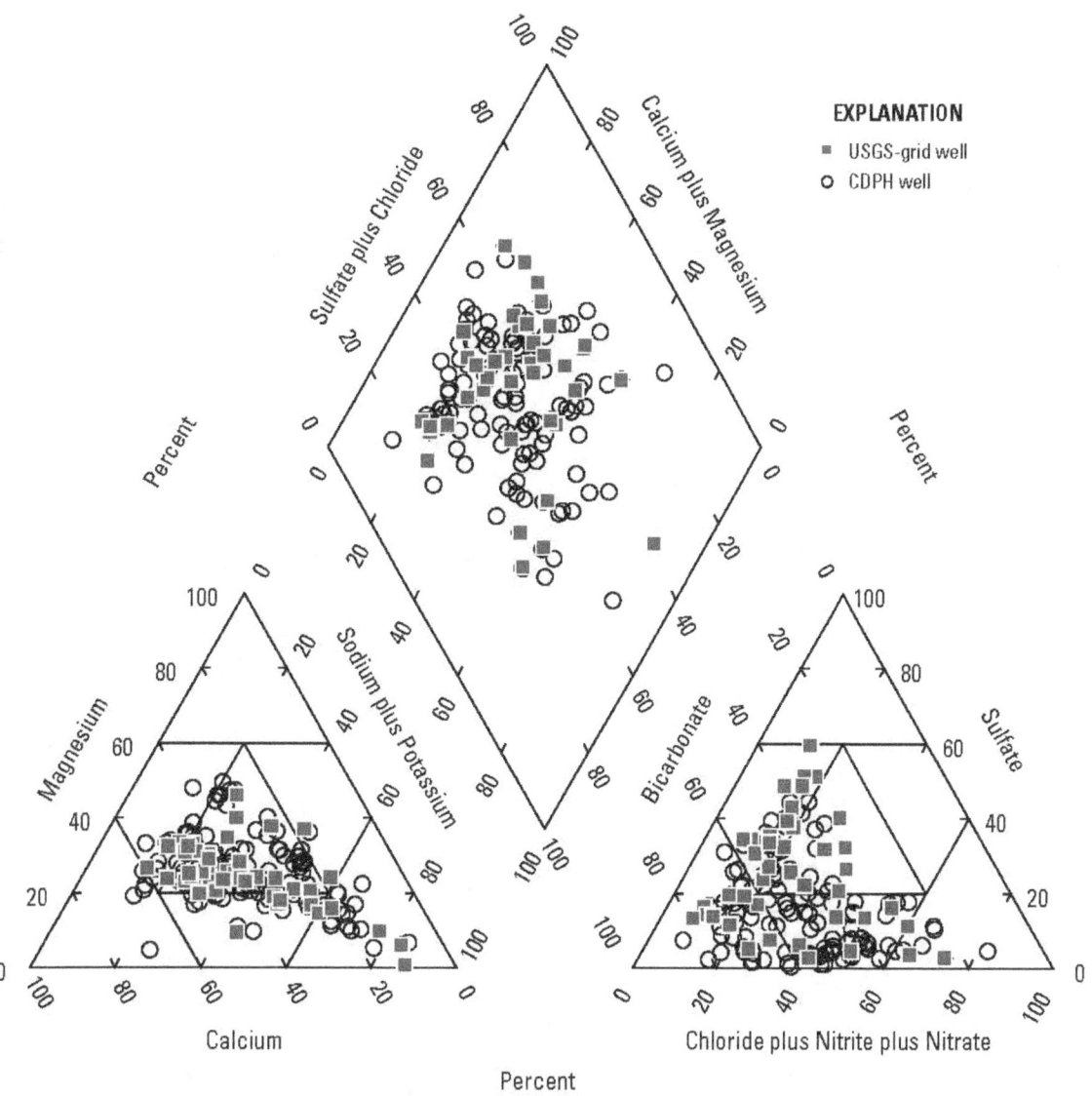

Figure E2. Selected inorganic data from USGS-grid wells and from all wells in the California Department of Public Health (CDPH) database that have a charge imbalance of less than 10 percent, Monterey Bay and Salinas Valley Basins study unit, California GAMA Priority Basin Project.